FUGITIVE ESSAYS,

UPON

INTERESTING AND USEFUL SUBJECTS,

RELATING TO

THE EARLY HISTORY OF OHIO,

ITS GEOLOGY AND AGRICULTURE,

WITH A

BIOGRAPHY OF THE FIRST SUCCESSFUL CONSTRUCTOR
OF STEAMBOATS;

A DISSERTATION UPON THE

ANTIQUITY OF THE MATERIAL UNIVERSE,

AND OTHER ARTICLES, BEING A REPRINT FROM
VARIOUS PERIODICALS OF THE DAY;

BY CHARLES WHITTLESEY,

OF THE LATE GEOLOGICAL CORPS OF OHIO AND OF THE UNITED STATES.

HUDSON, OHIO:
SAWYER, INGERSOLL AND CO.

1852.

WILLIAM H. SHAIN,
HUDSON STEREOTYPE FOUNDRY.

PENTAGON
STEAM PRESS.

PUBLISHER'S NOTICE.

THE articles which we now offer to the public, in book form, are selected from the numerous Essays written by Mr. WHITTLESEY, and published, during a series of years, in magazines and periodicals.

We have thought them too valuable to be lost, and believe that we are performing an acceptable service to the community in reproducing them at this time in a permanent form.

Many of the periodicals in which they appeared are but little known, having unfortunately, like most western monthlies, enjoyed only a brief existence and limited circulation.

Mr. WHITTLESEY'S liberal contributions to them have, although requiring much research and labor, added nothing to his purse, and very little to his reputation. We offer them to our readers for the solid information they contain.

Much of it belongs to our home history, and nearly all relates to the West as it is now, or has been since the advent of white men within its borders.

It is quite natural that a gentleman, reared in the western forests, and whose life has been spent in surveys and explorations throughout the territory northwest of the Ohio, should feel an ardent interest in every thing that relates to it, in natural or local history, statistics, agriculture, biography, or general condition.

These articles are given as first composed, not rewritten or retouched. Most of them were written in haste, in the interval of active field duties, and sent to press without revision.

A professional author would probably have polished his phrases, expunged and inserted words, and made some preparation to meet criticism, before permitting his works to appear before the public. The style is condensed and brief, with but little of that ornament which characterizes model literature; but those who read for the purpose of obtaining information will excuse this want of finish in recompense for the compact form in which that information is presented.

This selection is made from about as much more printed matter of a similar kind, published by the author, in a transient form, during the last fifteen years. An equal amount of the same cast is yet in manuscript, consisting of lectures delivered before literary associations, and never printed.

There may be discovered in the articles on early western history, some repetitions, which, however, could not be stricken out without injuring the whole piece.

If it shall appear that subjects are introduced, not in every respect agreeing with the *present* state of knowledge, the reader will remember, that it is many years since the series was commenced; and he can determine the *date* of each article by reference to the caption.

SCHEDULE OF SUBJECTS.

SCHEDULE OF SUBJECTS.

REVIEW

OF THE

TRANSACTIONS OF THE HISTORICAL SOCIETY OF OHIO,

VOLUME FIRST, PART SECOND.

[Hesperian, October 1839.]

THE appearance of this *second part* of the first volume will relieve the society from an unfavorable impression produced by the publication of part first, about a year since. We do not understand that every paper which may find its way to the files of the Society is therefore entitled to publication in a *permanent form;* but that the abiding works of the Society are intended to be solid, historical matter. Addresses pronounced at its meetings may, or may not, contain matter of that kind; and if it should be thought a courtesy, due the authors, to give them to the public in print, it may be done with great propriety in pamphlet form. The object of those compositions is, in general, more to amuse, and keep in action, the spirit of historical research, than to convey to us that kind of information. Of this character, or of a kind widely removed from the subject of history, are the addresses of the Hon. B. Tappan, Mr. J. H. James, Timothy Walker, Esq., and Gen. James I. Worthington, contained in the first volume. They are interesting papers, and highly creditable to the authors and the institution, as literary productions: but, being evidently written, not for the press, but for oral delivery, to instruct and please an audience, rather than to inform the youth of the West, the individual who opens this book, in the reasonable expectation of an historical feast, will be somewhat disappointed to find much of its space occupied by writings upon education, law and political institutions. It is not to the addresses or to their publication, nor to the custom of enlivening annual meetings with essays upon general subjects, that we object; for much of the interest and value of the Society is drawn from this practice.

The form and manner of publication hitherto adopted will, however, we fear, destroy the popularity and usefulness of its permanent works. Philosophy is a word associated with history, in the name of incorporation, and this embraces almost the whole range of human knowledge; but the main design has been, is, and ought to be, the preservation and publication of historical facts. The practice we are noticing is not of so much consequence at present, as it may be in future, when these precedents shall have made it a law.

It has been said that Ohio, in common with the new States, would furnish but a meager subject for the historian. One of this opinion must have reflected little upon the transactions of which the Mississippi Valley has been the scene, since the year 1673, when the Frenchman first made his appearance within its bounds. France, as intimated in the preface to these Transactions, is probably the repository of all the important records of this early period. To the inhabitants of that country, much more is known of the events of the region we inhabit, than by ourselves. It is to be recollected, English dominion was bounded in fact by the mountains, but the restless spirit of the Frenchman led him beyond this barrier, into that rich wild, now the seat of the power, wealth and resources of the Union. He sailed through the great lakes of the North, traversed the thick wilds of the West, floated with the current of our broad streams, built forts and opened a commerce with the Indians, while the Swiss, English and Dutch, scarcely penetrated beyond sight of the Atlantic. Improving upon the English scheme of western colonization, France and her enterprising citizens intended to take virtual possession of the North American continent. They effectually encircled the lodgements of their rivals, not only by a line of posts, the innermost of which were Ticonderoga, Stanwix, Niagara, Erie, Venango, Pittsburg, Loramies, Vincennes, Cahokia, and the mouth of the Ohio, but they formed stations through all the shores of the lakes, and the country between them and the Ohio and Mississippi; and, what is more than the mere occupancy of this new kingdom, they, by means of priests, presents, liquor and troops, brought the red man to terms of friendship. What can be more interesting than a full disclosure of those events? At the time of the French war of 1756, there were posts north of the Ohio, in what became, in 1787, the Northwestern Territory, at the following places: French Creek, Pa.; Du Quesne, at Pittsburg; Fort Sandusky, at Sandusky City; on Maumee; at Detroit (called Ponchartrain); Mackinaw; Fox River of Green Bay; St. Joseph's, mouth of St. Joseph's River; Lake Michigan; Crevecœur, and St. Louis; on the Illinois; mouth of Missouri; Cahokia; mouth of Ohio; Kaskaskia; Vincennes; mouth of Wabash, and mouth of Scioto. Occasionally a stray Englishman had crept over the Alleghanies and caught a prospect of the rich fields

1*

beyond; but until 1749 no general efforts were made by their sovereign, or his authorized agents, to occupy that country.

The charters of the colonies were broad enough, to be sure, and covered the Frenchman completely with a paper title. But the grant of a king three thousand miles off, or even the treaties of the six nations between 1684 and 1744, were little cared for by him. France had *actual possession* from Quebec to New Orleans; her citizens, her troops, her traders and missionaries threaded the woods and the streams of the Canadas and the Western States, from St. Lawrence to the Gulf of Mexico. Not a moment since this occupation commenced, to this hour, has the West ceased to be a place of interest. There are all the travels of Marquette, La Salle and their confederates, the dangers of the little establishments called forts, the hardships of artisans and farmers who clustered about them for protection, Indian kindnesses, murders and jealousies, upon which our citizens are still mainly uninformed. Then comes the enlargement of Gallic power and influence, till the British crown becomes uneasy, the efforts of the two powers to join the occupants of the soil to each party, both of whom were intent upon the destruction of their allies, the warning and suspicion of the Indian, the severe wars of the two rival nations upon a ground to which neither had a complete title, the success of British arms, and a thousand reminiscences of the past, belong to the territory we inhabit.

In 1756 came the contest of arms along a frontier line, which then lay within the present limits of Pennsylvania, New York and the New England States. This decided, ere long we read of the expedition of Col. Lewis and Lord Dunmore, west of the Ohio, in 1774, wherein the colonies fought the aborigines under the English rule; and soon we find the same troops engaged against the same foes, enlisted in the British ranks. Of the period between the domination of France and the close of the American Revolution, we have but little more knowledge than of the preceding. The line between historical light and darkness must be drawn near the date of the year 1783.

Are the exploits of the previous one hundred years unworthy of remembrance? We care not whether it is the record of the deeds of the Frenchman or the Englishman, the white or the red brother. These national distinctions do not impair our desire to know of their daring actions, nor personal antipathies prevent our admiration of their bravery and their enterprise. And can it be truly said, that the West is barren of materials for substantial history? Ohio does not embrace all the ground on which these interesting things occurred; but prior to the occupation of Marietta, in 1788, the progress of events, beyond the Alleghanies, was so connected with our territory, that without the whole any story would be incomplete.

As usual, the subject seems to attract more attention abroad than at home. The North American Review (Boston) has of late thrown open to our view many hidden sources of information, which it becomes us to turn to account. In the annual address of J. H. Perkins, Esq., 1838, it is stated that the manuscript journal of the first (or English) "Ohio Company," organized in 1748, is in existence, and in possession of the Hon. Charles F. Mercer, of Virginia. The author of the articles upon the French and English discoveries in the West, attaches in a note full references to his authorities, some of which we name:—Memoires Historiques Sur Louisianie, Paris, 1753, Present State of North America, London, 1755.

Pownall's Memorial, or Service in North America, 1756, London. The Contest in America, London, 1757. Boquet's Expedition, London, 1776. Charlevoix, La Hontan, Hennepin, Tonti, etc., Paris, prior to 1744. Plain Facts, etc., Philadelphia, 1781. Major Roger's Journal, and Concise Account, etc., London, 1765. Mr. Sparks, in the Life and Correspondence of Washington and Franklin, has performed the highest service to the cause of Western annals. His opportunity was a fortunate one, having access to volumes of old magazine manuscripts, and to public records in America, France and England.

Of the above works, some may be had in this country, and the remainder abroad, without incurring a very heavy expense. The papers relating to the Symmes' Purchase are said to be in existence. Those of the *second* "Ohio Company" are at Marietta, and the Connecticut Land Company at Hartford. Is not the collection of these precious relics of the early day a proper duty of the Historical Society of Ohio?

The narrative of Judge Burnet, which occupies one hundred and eighty pages of this *part* of the first volume, is of high interest. Since 1796, this astute and venerable man has lived in the city of Cincinnati. His acuteness of observation, tenacity of memory, and the practice of taking occasional notes, in connection with the fact of his long standing at the bar, and almost continual exercise of some public trust, municipal, legislative, or judicial, must give unusual value to these letters. Nothing could be more appropriate, both to the society and the individual. The latter has fulfilled a pleasing duty to the community through the proper channel; its execution, also, may be said to be equally happy with the other circumstances. The habits, manners, and weaknesses of the pioneers, are set forth with life and interest; and substantial matters relating to legislation, the organization and the progress of government, agreeably interspersed with anecdote and recital. We shall give the substance of some of the most characteristic passages. His acquaintance with the military of the West was of course intimate, of whom it is said, page 12, "The vices of idleness, drinking, and gambling, were carried to a greater extent in the army at that time than at any period since. A very large proportion of the officers of General Wayne's army were hard drinkers, General Harrison, Governor Clark, Colonel Shornberg, and a few others, being the only exceptions." These lamentable consequences are attributed to the absence of libraries, of men of refined taste and learning with whom to associate, yet more particularly to the want of accomplished female society. Since that period, much improvement has been observed in the moral character of our frontier posts. Education, a strong barrier to dissipation, has become universal among the commissioned officers; libraries are provided at almost every station, and the devoted wife accompanies her husband to those remote and forgotten spots, far up the Mississippi, the Missouri, and the Arkansas, to enliven the loneliness of his retreat.

Judge Burnet's opinion of the French traveller, Volney, who spent some time in Cincinnati and vicinity, in the fall of 1796, did not seem to be very exalted. "He was retiring, unsociable, and unusually credulous. Some officers, who travelled with him from this place to Detroit, availed themselves of this weakness, much to their amusement. One of the results of this play upon his *weakness* (no small thing in a traveller) was a conviction that the Ohio river in floods had been known to set back to the foot of the rapids, in a creek near Fort Greenville, now Greenville Court House, in Darke county, a point two hundred and twenty-two feet above Lake Erie, and three hundred and fifty-five feet higher than the river at Cincinnati."

Next follows a succinct account of the first legislative and judicial system of the North-western territory. Of this too much is known to need remark, being a system more thoroughly anti-republican than military rule itself. A just tribute is rendered to the memory of General Rufus Putnam of Marietta. A fine illustration of the phrase, "great effects flow from small causes," is recorded on page 17, in relation to the establishment of the city of Cincinnati. North Bend was the ground selected and surveyed for *the town* of the Symmes' Purchase, by the proprietor. It lies at the neck of the peninsula, between the great Miami and the Ohio. It appears in every view to have been a good location, and the present passage of this neck by a tunnel of the Whitewater canal, and the construction of this work thence twenty miles up stream, along the margin of the Ohio to reach Cincinnati, is evidence of the good judgment of Judge Symmes. The city of North Bend was laid out, and the troops from Fort Harmar landed at the place, for the purpose of erecting works. The commanding officer, somehow, became enamored of a black-eyed female, who, however, had a husband on the spot. To avoid consequences, the discreet lord removed to Cincinnati, taking the bright eyes of his wife along, which placed a half day's journey between them and the epaulets of the officer. The latter soon found the position of North Bend too weak for successful defence, and determined to reconnoitre the neighborhood of Cincinnati. The result was a confirmed opinion that the site of Fort Washington, somewhere in the vicinity of Mrs. Trollope's folly, on Third street, was a stronger military point than the heights of the peninsula at North Bend, and the forces were transferred thither. Protection thus withdrawn from the mouth of the Miami, its fate was sealed, and the destiny of the Queen City settled in a manner that, without the evidence before us, would have been considered fabulous.

As usual, in early settlements, one of Judge Lynch's courts of final jurisdiction was established in the colony. Patrick Grimes paid the penalty of the *law* provided for the case of stealing cucumbers, by receiving twenty-nine lashes on the naked back. But at the next sitting of the judge, the culprit disregarding process, fled to the garrison.

Mr. McMillan, who personated Judge Lynch, was next called upon, by a sergeant and two men, to attend upon the commandant of the fort. A pitched battle followed, lasting about twenty minutes, in which there were none killed on the spot; but four (all present) were seriously wounded. A Court of Quarter Sessions was soon organized. "In my early intercourse with the officers of General Wayne's army, I could not but feel surprised at the levity and calm indifference with which they spoke of exposures and hair-breadth escapes. I was certain that this did not proceed from any want of natural tenderness or sympathy. It seems to be a beneficent provision of nature, that men, who are timid, sensitive to danger, and disposed to sympathy, would cease to be influenced by such feelings, when duty brings them into scenes of peril and cruelty."

The difficulty between General Wilkinson and General Wayne was a cause of much disagreeable feeling in the army at the time, and each party had strong partisans. General Wilkinson went so far as to prefer charges against his Commander-in-Chief, and omitted few opportunities of degrading him with his officers. This notorious person is represented as a most fascinating and polished individual, calculated to attract friends to himself, and any cause he chose to advocate. Gen. Wayne's death on Lake Erie saved this agitator the ignominy of a defeat upon his own charges. A strong example of the exaltation of military feeling is related of Major Guion, the most uncompromising enemy of the General. News came of his decease. "What!" says he, "General Wayne dead! dead! then let enmity die with him."

We pass over much interesting matter to notice a statement of the author relative to some curious fossil stumps, or roots of trees, found beneath the surface, at Cincinnati. And to give our eastern friends a specimen of the veracity of writers and travellers, we insert the account given by Mr. Priest, in his "discoveries in the West," page 130. "In 1826, more than eighty feet under ground, there was found, on the banks of the Ohio (Cincinnati), the stump of a tree, three feet in diameter, and ten feet high, cut down with an axe, the blows of which are yet visible." The deductions of the author are, first, that it was antediluvian. Second, that the Ohio river did not exist before the flood. Third, America was peopled before the flood. Fourth, The antediluvian Americans knew the use of iron. Hear Judge Burnet upon the facts of the case; the inferences we leave to take care of themselves.

"The facts are simply these: in sinking a well in 1802, at the depth of ninety-three feet, I found two stumps, one about one foot, and the other eighteen inches in diameter, standing in the position in which they grew. Their roots were sound, and extended from them horizontally on every side. The tops were so decayed and mouldered that no opinion could be formed of the process by which the trunks had been severed from the bodies."

It is well known that the British did not fully execute the treaty of 1783, until the year 1796. They retained Mackinaw, Detroit, and Maumee, till after Wayne's victory over the Indians, at the Rapids, in 1794; and without the expedition, and the success which followed, resulting in the treaty of Greenville, it is doubtful when possession of those posts would have been given to us. The North-western Territory was divided into *four counties*, prior to the surrender, and their respective capitals were, Kaskaskia, Vincennes, Cincinnati, and Marietta.

The county of Wayne was erected soon after the delivery of Detroit, and the latter place became the seat of justice. Lawyers from Cincinnati practised at Marietta and Detroit, and sometimes at Vincennes. This circuit of the first three counties was regularly made by Judge Burnet, and his brethren of the bar, till 1803. A graphic description of the customs of Detroit will be found on page 50, quite equal to the style and manner of Washington Irving's Sketch of the Mackinaw Fur Traders. "Like men disposed to enjoy life, while it might be in their power to do so, they provided in great abundance the delicacies and luxuries of every climate, and as often as they returned from the cold regions of the North and West, to their families and comfortable homes, they did not spare them. No genteel stranger visited the place without an invitation to their houses, and their sumptuous tables; and, what is remarkable, they competed with each other for the honor of drinking the most and best wine, without being intoxicated themselves; and of having the greatest number of intoxicated guests." It appears that most of the British merchants of Detroit eventually crossed over to Sandwich and established themselves in business; but a friendly intercourse was continued. As evidence of this feeling, we have a full account in the celebration of the king's birth-day, in which the Americans joined. Afterwards, the members of the bar fulfilled an engagement, contracted on the spot, to spend a day and night at Malden. On these occasions, the English fashion of crowding wine upon their guests in profusion was not forgotten, and we are told that "although more wine was drank" than the writer had ever witnessed at such times, no animosity or bad feeling was excited. The British extolled George III. to their hearts' content; the President of the United States came next in order, and his name occasioned as copious draughts upon the tumbler as the king's. We know of a much stronger case of personal comity, which occurred during the last war, on the Ontario frontier, though it was with some difficulty the British officers persuaded themselves to swallow the compliment. Major Lomax had been sent from Sackett's Harbor, or that vicinity, to the British Head Quarters in Canada, with a flag. They received him very hospitably, but as a precaution, kept him very close, perhaps not more so than usual in such cases. The dinner and wine followed, of course, and with them abundance of toasts. An English officer, rather mellow with port, gave, President Madison, dead or alive." Major Lomax felt called upon to reply, and offered the health of the *Prince Regent*, "drunk or sober."

Judge Burnet's relation of the incidents of these journeys, from one court-house to another, three hundred miles distant, through a trackless forest, is often intensely interesting; but the reader must look to the work itself, we can retail but one. It would seem that Mr. St. Clair, son of Gen. St. Clair, the Governor and warrior, either through the influence of personal appearance, or official relationship, was much more caressed by the squaws of an Indian village, where they stopped, than the author. "An old wrinkle-faced squaw was extremely officious; her attentions, however, were principally confined to Mr. St. Clair; she kissed him once or twice, exclaiming, *you big man, Governor Son*, and turning to us, said, with some disdain, *you milish*."

A just tribute is paid to the character of General George Rogers Clarke, of Kentucky. "When I was induced to visit him by the veneration I felt for his talents and services, his health was much impaired by intemperance, but his majestic and dignified deportment, and strong features, bore the impress of an intelligent and resolute mind, and immediately brought to my recollection the personal appearance of Washington, to which it seemed to approximate."

In 1798, the North-western Territory contained five thousand inhabitants, and of right proceeded to establish the second grade of government. Of the legislative council of five, provided as advisers of the Governor, Judge Burnet was one during the continuances of this form of administration, or till 1803. Many of the early laws appear to have been drawn by him. The election of William Henry Harrison, as first Territorial Delegate to Congress, took place October 3d, 1799, by a majority of one vote. The next session of the legislature took place at Chillicothe, November 3d, 1800. During the sitting of the second General Assembly, a mob came together for the purpose of annoying the Governor (St. Clair), who seems to have been unpopular. This assemblage was renewed on a second night, and in consequence a law was passed restoring the seat of government to Cincinnati.

There are many points connected with the origin of our present government and constitution, not satisfactory to the author. He seems to consider the application for admission into the Union, before we attained a population of sixty thousand, and when, as a matter of right under the ordinance, we should have taken a stand upon the footing of the old States, as the cause of many evils. As it was in the power of Congress to grant or refuse the request, they imposed upon us terms which he considers unjust. In a subsequent address to the Society, this subject is amplified, and fully discussed. The principal objection is the relinquishment of the right of taxation on government lands, and for a period of five years after sale. The steps required by the ordinance and republican usage to compose a constitution were not followed, and the proceeding is considered illegal. It never came before the people, and the convention which framed it was ordered *by Congress*, and not the people, and so constituted as to make the instrument binding when completed by them.

Of Governor St. Clair, a few words must suffice. "He was plain and simple in his dress and equipage, frank and open in his manners, and accessible to persons of any rank. He retained a large share of popular favor till the close of the first session of the legislature. Soon after that body commenced its legislative functions, he exhibited a disposition to extend his power. The construction he gave to the ordinance was such as confined the will of the legislature within very narrow limits."

The ordinance giving him an unqualified veto, he considered himself as authorized and required to decide upon the expediency of all their acts. Of thirty bills, *eleven* were cut off in this way. This accidental state of things, occurring when the convention were in session, is thought to be the cause of their stripping the Executive of almost every respectable power, by the terms of the constitution.

"St. Clair was a man of superior talents, extensive information, and great uprightness of purpose. The course he pursued, though destruction to his own popularity, was the result of an honest exercise of judgment."

The object in examining a work of this kind is not so much to present the contents, or even its substance, as to carry a general idea of its merits, and if worthy of attention to induce its perusal.

We find it necessary, to avoid extreme length, in this instance, to pass by much important matter, and for the remainder merely touch at occasional points.

We are told, page 90, that Ohio led the way, in hallowing that memorable day, the anniversary of the Declaration, by recognizing the free enjoyment of personal liberty, through all its sacred forms, except in certain cases of crime.

Page 100 has an address of the legislature of 1799, to John Adams, President of the United States, embracing terms of strong compliment. The explanatory note discloses an important proposition, made to our commissioners at the treaty of Paris, 1782, not generally known.

The Ohio and Mississippi were insisted upon as our western boundary.

Dr. Franklin listened to the proposition, the Count de Vergennes favored it, and Mr. Adams at first stood alone in opposition to the measure, threatening to retire from the negociation. Mr. Jay soon sided with Mr. Adams, and Franklin finally concurred.

The history of some of the members of that legislature (1799) is given in brief. Judge Sibley of Detroit, Gen. Darlington of West Union, and Judge Burnet, are the only survivors. The most extended biographical notice is that of John Smith, afterwards a Senator in Congress, and finally implicated with Burr in his supposed conspiracy. Mr. Smith stated to the author, that his journey to Florida and Louisiana in 1806, was, by private request of Mr. Jefferson, to ascertain the feeling of the Spanish citizens and officers, in reference to the expected war with Spain. The prosecutions of the Government against him, in the next year, however, brought him to ruin.

In selecting members of the legislature in those days, "party influence was scarcely felt, and I can say with confidence, that since the establishment of the State Government, I have not seen a legislature containing such a large proportion of aged, intelligent, and discreet men." Pursuing the history of the formation of our State Government, we find, page 112, the reason of that novelty in judicature, a travelling court of *dernier resort.* The members of the convention could not decide upon the county or town in which the Supreme Court should be fixed, and to satisfy all sent them on horseback to every county in the State once a year.

Letter V, containing twenty-five pages, is mostly taken up with a novel discussion upon the right of the State to *tax Congress lands.* The author belonged to the minority upon the question of erecting a State Government, at the period of 1802, and still conceives that the people lost much by the terms of that admission. The excitement of those times upon this question was equal to that of any subsequent period, and the victors then, as now, exulted over their success. The remembrance of the strife of the occasion is not wholly effaced from the mind of the writer, who still expresses himself with some feeling. But it would be difficult to answer the argument advanced in support of our right to taxation over unsold government lands. It seems to be quite clear, that unless the terms of the relinquishment of the tax upon the *sales* of Congress lands, for the space of five years, includes the assent to relinquishment, or implies an acknowledgment of the non-existence of the tax right, the power is still vested in the State.

The seventh and last letter is occupied with desultory recollections of a highly instructive cast.

The project of constructing a canal around the Falls of the Ohio, on the Indiana side, was attempted, and some advance made towards its completion in 1817–18. In August, 1819, the river is stated to have been so low that its whole breadth at the Falls was only twenty-four feet, the water passing through a deep channel like a canal, with a division of rock in the center, and extending one-third of the length of the rapids. The old system of government sales upon credit, is shown to have threatened the ruin of western settlers; there being, in 1821, *twenty-two millions of dollars due the Government*, and an almost entire inability to meet it. The plan of allowing a relinquishment of the unpaid portions of the land in certain cases (proposed by Judge Burnet) finally prevailed in Congress, and the West was relieved. A short history of the canal donations of the General Government, and a brief notice of Simon Kenton, brings us to the close of these invaluable essays.

Taking so conspicuous a part as the author did in the doings of the Territory and the State, much that is personal necessarily occurs in the narration, but this is not the least interesting portion of its matter. We must pass over the remainder of the book, not of an historical character, and confine ourselves to those which are. The discourse of General Harrison upon the aboriginies of the valley of the Ohio occupies fifty-seven pages. In this case we are equally fortunate, in the fitness of the individual who undertakes to enlighten us. No person living has had as thorough acquaintance with the North-western Indians, as the gentleman whose name is just written. As a citizen and officer, in war and in peace, as a guest or a governor, in all conditions and circumstances, he has observed their anomalous character.

A portion of the discourse is taken up with a reference to the *ancient race* once occupying our valley in immense numbers, and whose habitations and temples still remain. Gen. Harrison supposes them to have been strictly agricultural; and most of these constructions dedicated to residence and religion; that the works on the river Ohio were of a different character, and were the result of necessity, intended as defences against a concerted invasion, from both the north and south, and that here they made resistance; but gradually retired down the river under defeat, making the last stand at a strong point near the mouth of the Great Miami. Of all speculations upon the design of these works, none are satisfactory to us, and none less so than those which give to them a military object. When a full collection and description of those interesting remains is obtained, perhaps some rational theory may be formed. At present, we have merely light enough to produce confusion. It would be easy to occupy several pages with reasons against the defensive character of these works, applicable as well to those on the Ohio, as elsewhere.*

But we hasten to consider the main discussion of this pamphlet; being the early history of the Indian occupants of Ohio, at the commencement of the white settlements; an occurrence which, it is worth remark, took place much later here than in the newer States west of us. They were composed of the Wyandots, Miamis, Shawanees, Delawares, a remnant of the Mo-

* See "Monuments of the Mississippi Valley," by E. G. Squier and E. H. Davis.—Smithsonian Contributions, vol. i. 1848.

higans connected with the Delawares, and a band of Ottowas.

Although the Six Nations claimed the north-eastern portion of this State, but few, if any, resided there. It is and has been disputed, whether the Six Nations ever conquered or occupied the country watered by the Scioto and Great Miami. Gen. Harrison says, their eastern boundary was certainly east of the Scioto, when the whites came to this country. Franklin, Clinton, and Colden, assert, and endeavor to prove, that the Iroquois once conquered and colonized even to the Mississippi. The English claim to the territory north-west of the Ohio, as opposed to the French, rests upon a grant of the Six Nations, as early as 1684. A profound antiquary, in an article upon the English discoveries in the West, North American Review, July 1839, concurs with the statements of De Witt Clinton and others, that they had, prior to 1680, overrun most of the modern north-western Territory. Gen. Harrison opposes this opinion, and does not admit that they ever possessed lands west of the Scioto. It is an interesting examination, beyond our limits to transcribe. We merely improve the opportunity to add an item of evidence in favor of Gen. Harrison's belief. In 1796, when the agents of the Connecticut Land Company proceeded to make surveys of the Western Reserve, the Indian title was not fully secured. Gen. Cleveland held a council with the Six Nations, or a part of them at Buffalo, for the purpose of taking quiet possession. At this time they claimed nothing beyond the Cuyahoga and Tuscorora rivers, and the "old Portage path," or Portage, connecting these streams across the Akron summit. They considered this line as the boundary between them and the western Indians, and gave no rights beyond the Cuyahoga. These streams and the old Portage path were used in common by the Indians on both sides, for transportation, of which so much took place, that the trail or path, from one to the other, is still visible. The company took possession of the land east of the agreed boundary, and surveyed to it; the Indians on the west of it occupying their side, and not molesting the occupation of the whites. It was not until 1806 that the Land Company, or Congress, obtained full possession from the Indians of the western shore of the Cuyahoga.

The discourse locates the different tribes at about 1650, as follows: the Iroquois confederacy remained in their original position between Labrador and the Delaware, or great Lenape nation, whose northern limit was somewhere in southern Pennsylvania. The Wyandots or Hurons occupied both shores of western Lake Erie, and extended southward to the Ohio. The Miami confederacy, the most powerful of the Indian combinations, lay along the Ohio, from the Scioto westward, around Lake Michigan to the Mississippi.

The Iroquois fought and conquered the Delawares on the south, took possession of their land, and forced them to assume the name of women. Some time during the seventeenth century, those warlike and generous tribes moved upon the Wyandots, and defeated them. This battle is said to have been fought in canoes, upon the waters of Lake Erie, and great fatality resulted. The Wyandots withdrew westward *for a time*, but returned again in the eighteenth century, when it is *probable* the Cuyahoga was mutually agreed upon as their eastern border.

The Miamis, possessing most of western Ohio, are thought never to have been at war with the Six Nations, or at least never to have been conquered by them. With the Cherokees and Chickasaws they were ever at war. Gen. Harrison considers Sandusky as the western limit of occupation by the Iroquois, and that possession a temporary one. The Shawanese Indians were emigrants from Georgia and Florida, within an hundred years. They first came to the country of the Miamis, low down on the Ohio, and afterwards moved to the Scioto. Black Hoof, their chief, died not long since, and he was *born* in Florida.

The Indians engaged against the United States in open war, from 1790 to the peace of 1794, were the Wyandots, Delawares, Shawanese, Chippawas, Ottowas, Potawatamies, Miamis, Eel river Indians, and the Weas. *Three thousand* warriors constituted their strength at this time; while the Miamis alone could have mustered that number a short time before. As late as 1793, they had determined to have the Ohio as their boundary. The battle of the Rapids, a year after, forced them to the Greenville line, and for many years awed them into quietness.

We are compelled to take leave of this elegant and instructive production, and recommend its style and contents to the perusal of every western man. It was our intention to treat of the historical parts of the address of Mr. James H. Perkins. Those portions relate to the remote doings of the French, and the early occupancy of the English, of which we have given some slight outline in the commencement of this article. Mr. Perkins can not fail to be read through by every reader of any portion of this work. An address delivered at Marietta, on the forty-eighth anniversary of the settlement of that place, closes the volume. It is confined to the transactions of that region, and filled with matter of great value. Arius Nye, Esq., is the author, a gentleman bred within sight of the "Campus Martius," of the first settlement in Ohio. This post became a prominent point of attack, by the hostile Indians. Its history thrills with interest, but the most exalted sentiments connected with its recital arise on the consideration of the nature of those men who first broke in upon the forest-world of the West, and successfully planted civilization in the midst of the fiercest barbarism. Their like is never to be known again. In the progress and mutations of human affairs, such a concourse of circumstances will never arise. There can never be another such revolution as that of 1776. If that was possible, will there be again such patriots, such men? Then came the weakness of their country, and their own impoverishment; afterwards the offer of western lands, in compensation for military service, but requiring the protection of military force. The never-lessening patience, perseverance, and piety of those stern characters, has no parallel. With all these traits we behold the hourly exercise of courage, the cool contemplation of danger, acuteness of design, and vigor of execution.

In dismissing this work, we must express our extreme regret that no index or even contents of chapters can be found on its pages.

JUSTICE TO THE MEMORY OF JOHN FITCH,

Who in 1785 invented a steam engine and steamboat—planned, constructed, and put in operation the steamboat "Perseverance," of sixty tons, moving at the rate of eight miles an hour, in 1788.

[Western Literary Review, Feb. 1844.]

UPON the urgent request of John F. Watson, Esq., of Germantown, Pa., the writer about two years since undertook an examination of the claims of John Fitch, of Philadelphia, as an inventor and improver of steam engines and steamboats. I had long been of the opinion, which now pervades the civilized world, that mankind owed to Robert Fulton, not only the improvement but the *invention* of steamboats. And to ward off a smile, which those well versed in the history of the steam engine and its applications may be inclined to indulge at my expense, the following quotations are made. They show that, however erroneous my opinions were in regard to the real author of that contrivance which now rides on every navigable river inhabited by civilized man—defying the elements and exacting the admiration of all—there are very respectable persons entertaining the same opinions.

In 1838, Judge Tannehill writes: "When Fulton *first conceived* the idea of navigating our rapid stream with boats propelled by steam,"* &c.

Judge Story, in charging the jury in the case of Washburn & Brown *vs.* Gould, Boston, 1844, remarks: "Next after Fulton's *wonderful invention* of the steamboat, whose incalculable benefits," &c.

Still later, a writer in the American Review for January, 1845, says: "A few years more saw the spirit of Fulton arise, and *call into existence* what has proved perhaps the most important of the manifold agencies of steam."†

It will be impracticable, in the space at our command, to return through the space of *one hundred and seven years* to the days of the *Hulls of England*, who in the year 1737 obtained a patent "for a new invented machine for carrying vessels out of or into any harbor, port, or river, against wind and tide, or in a calm," with wheels at the sides of the vessels, and buckets on the periphery, and exhibit to our readers

* Hesperian, vol. ii, p. 106. † Am. Review, Art., Steam Navigation, p. 22.

the various contrivances in the nature of self-moving craft that have been thought of by mechanics.

It is not clear that the Hulls had a steam engine in contemplation as a moving power, although a huge chimney arises from the deck of their vessel which pours forth abundance of smoke.

Solomon De Caus had written of the power of steam more than one hundred years before this patent was obtained from George II., and from 1612 to 1623 had experimented upon its forces and application.

In 1663 the Marquis of Worcester, an Englishman, wrote of his hundred inventions, and mentioned "an admirable and forcible way to drive up water by fire."

Dennis Papin, like De Caus, a native of France, had lived and experimented upon steam in 1698; Savary in 1698; and Newcomen and Calley, or Cawley, had constructed a rude but improved engine in 1711. But James Watt, though living, had arrived at the age of only one year when Jonathan Hull applied for his patent; and the engine was sleeping in the philosophical rooms at the University of Glasgow, an awkward and cumbrous thing. It had hitherto been applied merely as a lifter, acting in a right line. It had been taught to open and shut some of the valves of the condenser; yet its energy was communicated only in one direction, relying upon the atmosphere to bring the piston back to its place. If the Hulls had at this early day an idea of an engine with a *rotary motion fitted to revolving wheels*, immortal honors are due to their names. The proof of this knowledge is so far deficient as to preclude this idea.

It appears to be a well-established fact, that John Bournouilli, in France, described a method of propelling a boat in water by means of a pump, and his conception was made public before the French Academy in 1753. The power of steam was not relied upon to work the pump, but that of animals or men. Even in the days of Papin and Savary, a boat was worked by animals on the Thames, which had wheels at the side, and was constructed by Prince Rupert. Some accounts, ascending to the time of the Punic war, mention a boat moved by oxen, which the Romans used for the transportation of soldiers across an arm of the sea.

The Spanish nation has lately produced a manuscript, said to have been found at Barcelona, and according to it, Blasco De Garay gave the velocity of a league an hour to a vessel of 209 tons, on the 17th of June, 1543, in the presence of Charles V. and his cabinet.

But without resorting to fabulous or uncertain stories, we continue our abstract of the progress of the *development of the idea* of steam navigation.

Next after Bernouilli came Genevois, in 1759; the Count D'Auxirron, of France, in 1774; the elder Perrier, also a Frenchman, in 1775. The Marquis De Jouffioy turned his attention that way in 1778 and 1781; built a boat to be moved by steam at Bourne le Dimes, which he described in 1783 to the Academy.

Jouffroy experimented many years, and his model boats obtained a considerable velocity.

In America, Oliver Evans states that he reflected upon steam vessels as early as 1772–3, but made no public declaration of his views. Mr. Henry, of Lancaster, and Andrew Ellicot, as appears from conversations with Mr. Fitch, had, during the progress of the American Revolution, secretly conceived of some plans for effecting the same object.

In France, the Abbe Raynal had projects of the same kind in 1781.

This rapid view of inventors and improvers in

steam power brings us to the time when experiments began to be made upon the same subject in America. The first with which we are acquainted in this country took place in secret, in presence of a few friends near Shepherdstown, Va., during the fall of 1784.* It was made by James Rumsey, a native of Maryland, and resident of Virginia, who had conceived of the project in 1783. Rumsey's boat had a capacity of six tons, and was first set in motion privately during the darkness of night, the first public experiment having been made in the year 1786 † or 1787. ‡

Mr. Fitch conceived of a plan to move water craft in April, 1785. Returning one Sunday from church, in the township of Warminster, Buck's county, Pa., a *chair*, a riding vehicle with wheels, passed along the road. Reflecting upon its motion, he supposed that it might be made to traverse the country by the force of steam. After a short time he concluded this to be impracticable, and turned his thoughts upon a scheme of propelling vessels in water by the same agency.

I deem it unnecessary to proceed farther in the chronological notice of improvements in steamboats. The reference to inventors will be made hereafter without much regard to the order of time.

There were in Scotland, Miller, of Dalwinston, 1787; Lord Stanhope, 1793; and Hunter and Dickinson, 1801; in France, Des Blanes, 1802; in America, John Stevens, Jun., 1790–1, and R. R. Livingston, 1798. This crowd of inventors labored, experimented, and suffered before Mr. Fulton's boat was constructed at Plombieres, in 1803. Their several machines had

* Reports of Congress Committees, 1836–7, Vol. ii. No. 317. Letters of Washington, Jan. 31, 1786, and Nov. 22. 1787.

† Reports of Committees, 1838–9, No. 265; and 1836–7, No. 317.

‡ Virginia Gazette of Nov. or Dec., 1787.

produced a speed of two or three miles an hour, which was not of sufficient practical importance to attract the attention of commercial men. But these facts show incontestably how little room there was for the exercise of *original invention* in the conception of a steamboat in the year 1793. They also show how little had been done towards perfecting the boat in the year 1783–5, when the American improvers took up the subject. My object is to establish some points wherein it appears that public sentiment has been misled from the truth. The passages above quoted, and numberless expressions uttered in conversation, whenever the *subject* of the achievements of steam is introduced; in fact, the universal sentiment of mankind has placed the era of steam navigation in the year 1807, when the *Clermont* made her first trip along the Hudson, at the rate of *four and seven-tenths* miles per hour.

This opinion so well fixed and so prevalent has descended from the generation which was astonished by the appearance of this boat, and of the *Paragon*, *Car of Neptune*, and their consorts, to another who have inherited the belief of their ancestors. Opinions derived from our forefathers require but little confirmatory proof. They are often received, not only without evidence, but without the desire of it; without research, without question. They amount to more than belief. They may be regarded as a prejudice, a condition of mind where contradiction is disagreeable, and of investigation not only disagreeable but forbidden. I have felt this bias, and can appreciate its force. My investigations have however driven me, step by step, from my original conclusions in regard to the date of the great impulse which inland navigation received by steam. A thousand circumstances have conspired to conceal from the public experiments made and results obtained at Philadelphia in the last century. An examination of those experiments may be tedious, but may also be interesting. It will introduce, in a new light, a man of extraordinary character, a man without education or property, struggling against adversity, against ridicule, neglect, want, and an accumulation of misfortunes, to perfect an invention which his age could not comprehend.

The investigation will exhibit a mechanical genius, so far absorbed in its great idea as to abandon self, suffer wrong, and perform astonishing labors, to accomplish a stupendous undertaking, in advance of the times in which he lived. I think it will be acknowledged, that the undertaking *was* accomplished. If the *improvement* of an engine to work on land upon fixed machinery was an act of sufficient merit to immortalize the name of James Watt, in England, will not the *construction* of an engine by John Fitch, in America, at the same time, and capable of driving a vessel upon water, ensure *his* name a place among those who have acquired reputation and honor as improvers upon Savary and Newcomen? How many have added valuable parts to the engine since it came from the hands of De Caus? Two centuries and a quarter of time have been expended in perfecting, increment by increment, a machine which, at this day, possesses the manual facilities of man, and the strength of many thousands. One man, one age, was insufficient to produce this result. The ingenious of many generations have spent the force of their mechanical talents in the study and improvement of the engine alone. Hundreds and thousands have wasted life and patrimony, during the last three quarters of a century, upon the improvement of the boat to be moved by its expanding powers. All this sacrifice, this thought, this loss of ease, peace, money, friends, and human life, was necessary to and worthy of the result. Every individual who made an advance upon his predecessors has been added to the list of fame, except the obliterated and forgotten name of John Fitch. How many of all those who ride over the waters of the earth, impelled by the force of steam, ever heard it pronounced? What boat of all the seas and rivers of the two continents has his name upon her sides? And among the scattered reminiscences which his surviving friends here and there publish to the world, how few are credited or even received in memory. Among those who have charge of the engine, now performing its delicate and wonderful duties in every city, town, and village of importance within the range of manufacturing enterprise, can *ten men* be found who are aware that John Fitch lived, and obtained the first regular rotary movement of its part in America? Oliver Evans, probably the most extensive improver of the engine in the new world, is well known to his countrymen. His name is familiar, and his exalted merits as a mechanic are acknowledged, not only in the United States, but on the other shore of the Atlantic. But from the imperfect biographical notices within our reach, it does not appear that in 1785 Mr. Evans had *constructed* any of his projected improvements upon engines and steam carriages.

Before Mr. Fitch had devoted his talents to the subject, Mr. Evans undoubtedly arrived at conclusions in regard to steam machinery; for in 1781 he had asserted that by the "power of steam he could drive anything, wagons, mills, or vessels." But if any of these conceptions were, at this time, reduced to practice, and made visible to the world, his historians are not in possession of the facts. His petition to the Legislature of Pennsylvania for a monopoly is dated in the year 1786. At this time, Mr. Fitch had constructed a working model with a cylinder, at first of one inch in diameter, and afterwards of three inches, in which *continuous rotary motion* was effected. Of his first attempts at an engine, he makes the following assertions: "What I am now to inform you of, I know will not be to my credit, but so long as it is the truth I will insert it, viz: that I did not know there was a steam engine on earth when I proposed to gain force by steam. I leave my first draft and descriptions behind, that you may judge whether I am sincere or not. A short time after drawing my first draft for a boat, I was amazingly chagrined to find at Parson Irwin's, in Bucks county, a drawing of a steam engine, but it had the effect to establish me in my principles, as my doubts lay at that time in the engine only."*

This was in April, 1785. The improvements of Watt were some of them patented in 1769; but his patents for rotary motion, four in number, bear date from 1782 to 1785. For parallel motion he received a patent in 1784. If Mr. Rumsey had at this time constructed an engine, it was such an one as, without rotary motion, would work a pump by a right line movement. For locomotives, for machinery in general, and for vessels, this continuous regular revolving apparatus is not only necessary, but indispensable. The uses of the engine, without the faculty of unceasing revolution, would be limited and insignificant, as they were before the time of Watt.

Mr. Fitch constructed a rotary engine and put it in operation. Did he invent as well as construct this rotation? Had the improvements made in England within the three previous years reached him in Pennsylvania? Were they generally known even in England at that time? were they published? were they sent to America? Had Evans communicated with

* Fitch's Manuscript Writings.

Fitch? Had Rumsey? and if he had, was there anything in the pumping engine that could be applied to the revolving engine? To sustain the affirmative of these queries, we know of no proof. The American colonies were at war with England, and intercourse was unfrequent until after the peace of 1783. The presumption is in favor of the negative to them all. Rumsey's machine was kept a secret until within a few months of April, 1785. Communication was irregular and slow, mails and newspapers few, and not within the means of poor mechanics like Fitch and Rumsey. Fitch himself was in the wilds of Ohio and the northwest, a prisoner among savages from 1782 to 1783, and from 1783 to 1785 remained in the obscure village of Warminster, laboring at the business of a silversmith to procure the bread of life.

In addition to these circumstances, there is the following paper, which may be found in the American State Papers (miscellaneous), vol. i. page 12. It is a copy of his petition to Congress, dated July 2d, 1790, and says, "That the great length of time and vast resources of money expended in bringing the scheme to perfection have been wholly occasioned by his total ignorance of the *steam engine*, a perfect knowledge of which has not been acquired without an infinite number of fruitless experiments, for not a person could be found who was acquainted with the *new engine* of Bolton and Watt. Whether your petitioner's engine is similar to those in England or not, he is at this moment totally ignorant, but is happy to inform Congress that he is now able to *make a complete steam engine*, which, in *its effects*, is equal to the best in Europe."

In 1803, thirteen years after this affirmation was made, the workshops of America could not furnish an Engine.* Mr. Fitch regarded the construction of the

* Renwick on Steam, 256.

engine as the *great obstacle*, and to overcome this what resources of knowledge were within his reach, other than the *original* inventive powers of his own mind? Dr. William Thornton, one of the Philadelphia company, says they labored "under the disadvantages of never having seen an engine, and not having a single engineer in our company, or pay, we made engineers of common blacksmiths,"* &c. In England, from whence alone such information could have been drawn, no such engine existed, and therefore could not have furnished the ideas, model, or descriptions for this. An engine for a cotton mill was incapable of driving a boat, without material alterations. Twenty years afterwards Mr. Fulton regarded the *adaptation of the engine* of Watt, made to order in England, expressly for the use of the *Clermont* upon the Hudson, as the greatest difficulty in his path to success. The new parts necessary to be added, the new form to be given to parts then in existence, in order to conform to the limited space of the hold, and to connect with the shafts or axles of the wheels, were to him the principal causes of doubt. At this time (1807), when the genius of Fulton was so much embarrassed by the *fitting and application* of the engine, its construction had been brought to a high state of perfection in Europe, and he had the assistance of Watt himself in overcoming those obstacles. In 1785–6–7, Fitch had no Watt, no engine, no machinist who had made or ever seen one, no workshop and tools, no detailed descriptions.

My purpose has been thus far to exhibit the powerful resources of his intellect as applied to *engines*. As we advance it will appear more fully what kind of an engine he produced, and what where its effects. For the present, let it be remembered, that the first *rotary*

* Lives of Eminent Mechanics.

engine in America was conceived by him in 1785, and a working model finished in 1786. I shall not dwell longer upon the proofs and arguments in favor of his *originality* as an inventor of this engine, leaving the facts advanced to produce their effects. If it is claimed that its prior existence in Europe takes away the merit of *first* invention, it can not be asserted that *the engine as applied to his boat* is not a *new* as well as an original invention. The moving agent of Jouffroy, if in principal, form, or substance, similar to the one of Fitch, not being known as such in America, in England, or in France, either in 1785 or in 1845, could not have been the model from which the Philadelphia mechanic obtained his ideas.

It will be necessary to omit the personal history of Mr. Fitch. His life, as a mere popular narrative, might be invested with an intense interest. It would prove the truth of his own words concerning himself, as "one of the most singular as well as one of the most unfortunate of men." It would exhibit misfortune and suffering, great undertakings followed by disappointment and despondency, great force of character, courage, and pride. It would show personal antipathies to be regretted and noble sentiments to be admired. Tenderness as a parent would appear in strange contrast with an unforgiving temper as a husband—honorable impulses and raging passions contending in the same bosom. His intellect bore the stamp of originality and independence, his foresight had the air of prophecy. He was born at Windsor, Conn., Jan. 21, 1743 (O. S.), and remained in that vicinity until about twenty-five years of age, receiving a very scanty common school education. By the severity of his father and elder brother, his life at home was rendered wretched; and he was bound apprentice to a watchmaker, without, however, having acquired a knowledge of the trade. An unfortunate marriage crowned the misery of his condition, and in 1769 he became the adventurer of fortune. After many wanderings, he became a resident watchmaker at Trenton, New Jersey, where he exercised his trade at the commencement of the Revolution. The demand for arms induced him to undertake the business of a gun-smith for the American forces, which exposed his property to destruction, when the British entered that village, in December, 1776. He joined the troops of New Jersey and endured the rigors of a winter camp at Valley Forge.

Retiring from camp, he recommenced the trade of a silversmith, in Bucks county, Pa., occasionally traversing the county on foot to repair the clocks and watches of the inhabitants. Having procured an appointment as deputy surveyor from the State of Virginia, he started for Kentucky with a knapsack upon his back and a compass in his hand, in the spring of 1780.

In the fall of 1781, he returned to Philadelphia, having made extensive surveys between the Kentucky and Green rivers. In the spring of 1782, collecting the fragments of $4000 which had been received in continental money, he was barely able to raise £150 Pennsylvania currency as a capital for Western adventure. At the mouth of the Muskingum this remnant of his fortune, invested in flour and goods, was captured and destroyed by Indians, two of the party killed, and nine taken prisoners. Fitch had the address to conciliate Capt. Buffaloe, the leader of the band, and the physical endurance to sustain the rigors of Indian slavery.

After various adventures he reached Warminster in the winter of 1782–3, penniless and dejected. Here he resided when the "unfortunate" inspiration came across his mind in regard to steam.

"From that time (1785) I have," he says, "pursued the idea to this day (1792) with unremitted assiduity, yet do frankly confess that it has been the most imprudent scheme that I ever engaged in. The perplexities and embarrassments through which it has caused me to wade, far exceeded anything that the common course of life ever presented to my view."

To comprehend more clearly what ground Mr. Fitch occupied as an inventor we must revert to the engine and the boat. An engine having long been known, it was not the subject of *conception*, as a new and unheard of thing. The idea of a boat to be moved by steam having been known and discussed, could not at that day have been thought of as a first invention. The proof, it appears to me, is conclusive that, in regard both to the engine and boat, Mr. Fitch had formed a connected plan before it was known to him that they existed; that he was therefore an *original*, but not a *new* or first inventor.

The field then open to the speculator in boats and engines embraced only the *improvements* and *applications*. The knowledge of this fact struck down his rising hopes of renown and usefulness. He had seen the rivers of the West, and had heard from the Indians of many lakes and streams far away in the north and north-west. He had trod the rich soil of the Kentucky, the Sciota, and the Miami rivers, and foresaw the capacity of the regions watered by those streams for the support of life and commerce. His brain was fired with the thought that the navigation of those rivers might be effected by his agency, and for a time luxuriated in the delusion that no mortal had conceived of a similar project. But stripped of those lofty anticipations, he persevered as an improver, in what he commenced as an inventor. What ground was yet unoccupied? The experiments of Rumsey,

4

though unknown to Fitch, had been made. A boat had been moved upon the Potomac, with a velocity of three miles an hour, and afterwards increased to "four or five." It is said that Jouffroy had procured a motion of three miles an hour likewise, unknown to Fitch. The abstract question of *a boat* was no longer open to invention, nor was the first construction of a *self-moving craft* to be made twice.

There was room, however, for the *invention* and *construction of such a boat* as should, by *different* arrangements, machinery and apparatus, *be an improvement* upon all known steam vessels.

And it is not of so much importance in estimating the value of an invention, to know all that has been done, as to regard what has been thought of or made which is identical. The fact that the end to be attained is the *same*, by no means precludes the idea of arriving at it by *different* methods. That meritorious mechanic Rumsey, selected *one* mode, Jouffroy *another*, and Fitch a *third*. The name of all these contrivances was the same, and the object the same. In each case there was a floating vessel and an internal power, the whole called by a common name, the "Steamboat." Between the central force and the water, or medium on which it must be applied, there was a gap not filled. The boat of our day has received a connecting apparatus, which transfers the power of the steam cylinder to the surface of the water by means of cranks, shafts, wheels, and buckets.

It is *this boat as reduced to practice*, that has received so much admiration, and been of such vast utility. At that day it did not exist, now it does; *who made it what it is?* The discussion is naturally confined to *such a boat*, and does not range through all manner of boats that have ever been thought of or tried.

To compare more directly the methods adopted by the two American competitors, and to compress the presentation of the subject as much as possible, I have postponed the description of these respective boats, and here introduce them side by side:

Description of the boat of James Rumsey.

"Rumsey's boat was about fifty feet in length; was propelled by a pump worked by a steam engine, which forced a quantity of water up through the keel; the valve was then shut by the return stroke, which at the same time forced the water through a channel or pipe, a few inches square, lying above and parallel to the kelson out at the stern under the rudder. The impetus of this water forcing the square channel against the exterior water acted as an unfailing power upon the vessel. The reaction of the effluent water propelled her at the rate above mentioned (four or five miles an hour), when loaded with three tons, in addition to the weight of her engine, or about one-third of a ton. The boiler was quite a curiosity, holding no more than five gallons of water, and needing only a pint at a time. The whole machinery did not occupy a space greater than that required for four barrels of flour. The fuel consumed was not more than four to six bushels of coal in twelve hours. Rumsey's other project was to apply the power of a steam engine to long poles, which were to reach the bottom of the river, and by that means push a boat against the current."*

Description of the boat of John Fitch,

Published by himself in the Columbian Magazine, Dec. 1788.

"The cylinder is to be horizontal, and the steam to work with equal force at both ends. The mode by

* Stuart's Anecdotes of the Steam Engine.

which we obtain a vacuum is, we believe, entirely new, as is also the method of letting the water into it, and throwing it off against the atmosphere without any friction. It is expected that the cylinder, which is twelve inches in diameter, will move with a clear force of eleven or twelve hundred weight after the frictions are deducted; this force to be directed against a wheel eighteen inches in diameter. The piston is to move about three feet, and each vibration of it is to give the axis (or shaft) forty revolutions. Each revolution of the axis moves *twelve oars* or paddles, five and a half feet. They work perpendicularly, and are represented by the strokes of the paddle of a canoe; as six of the paddles are raised from the water, six more are entered (three on a side), and the two sets of paddles make their strokes of about eleven feet at each revolution. The cranks of the axis set upon the paddles, about one-third of their length from their lower ends, on which part of the oar the whole force of the axis is applied. The engine is placed in the bottom of the boat about one-third from the stern, and both the action and reaction turn the wheel the same way."

Such are the two schemes of navigation which occupied the minds of Rumsey and Fitch, which became the ruling passion of their souls, and continued to employ their mental and physical energies until the last hours of life. It was in respect to these projects, so widely different, that between them an animated personal controversy arose during their lives for the honor of the priority and value of their respective machines. It would seem that they were pursuing paths so far asunder that no collision of this kind was necessary. It would be difficult to point out a single article common to the two boats, except the boiler and the floating craft. Whatever merit belongs to either may be shared in its fullest extent without derogating from the other. Rumsey invented in 1783, experimented in 1784, and more fully in 1786 or 1787. Fitch invented in 1785, experimented in 1786, and performed on a large scale in 1788. Both began as *original* inventors, but both had the mortification to learn that they were not *prior* inventors. For Rumsey's propelling arrangement, Bernoulli had pre-occupied the ground *fifty years* before, and as to the engine, the invention had been known more than a century. In this respect, however, the engine of Fitch is to the engine as an *improved* machine what Watt is to Newcomen. The latter was without circular motion, without, therefore, a complicated connecting series, without axles or shaft crank or cogwheels. The engine of the former appears to have included the "double-acting" principle, by which the old *atmospheric* engine was made a true and real *steam* engine, working, as he says, "with equal force at both ends."

This improvement, like that of a good rotary motion, was likewise introduced by Watt about the same time. We have no account of the precise time when Mr. Fitch made this addition, or whether it was original with him or not. Neither am I able at this time to ascertain the exact period when Mr. Watt first made known to the world his invention of the double-acting cylinder. It is reported that he retained it as a secret a long time after his own conceptions were matured. If made public in 1785–6, whether it had reached Mr. Fitch, at Warminster, is also involved in doubt. For this, as for the rotary apparatus, I refer to his own statement made, many years after, to the American Congress. His first model was completed in brass in August, 1785, having its machinery perfect, and bearing at the sides wheels instead of *paddles*. The paddles were adopted in 1786, after experimenting upon the wheels.

4*

Dr. Franklin, in the fall of 1785, had written an essay upon navigation, discarding the use of the wheels, and adopting the plan of Bernoulli.

The buckets of the wheels were found to labor too much in the water, entering, as they did, at a considerable angle, and departing at the same. They lost power by striking the surface, and afterwards by lifting themselves out of the water. This led to the substitution of oars or paddles, which entered almost perpendicular, and left the water inclined a little towards the stern. The construction of such a boat became to Fitch the highest object of his ambition. He applied to the Continental Congress for aid, representing the immense advantage its success would be to the western lands lately conquered by the American arms. He petitioned the legislature of Pennsylvania for money, representing in high wrought figure of the imagination the splendid consequences of the project, if carried into effect. He portrayed in the private ear of the Western and Virginia members of Congress, the achievements in reserve for steam through the agency of his contrivance. He wrote to Franklin in October, 1786, affirming the practicability of sea navigation by steam vessels, and every where, and at all times, boldly asserted as a prediction what we observe as facts. But none of his fervid representations produced the money, and he acquired the reputation of an insane man. Finally, by the construction, engraving, and sale of a map of the north-western territory, all of which was done with his own hands, in the workshop of his friend Cobe Scout of Warminster, and the impression taken in a cider press, he raised about $800, and in February, 1787, formed a company of forty shares, and commenced a boat of sixty tons. After innumerable vexations and delays, principally occasioned by the formation of the engine, the boat was put in motion, and made only *three miles* an hour. The machinery was so rough that the expected power of a cylinder of twelve inches was not realized. The company was discouraged, but another rally was effected, the shares doubled, and the improvements commenced. "I was," says Dr. Thornton, "among the number (shareholders), and in less than twelve months we were ready for the experiment."

The day was appointed. A mile was measured on Front street (now Water street), Philadelphia, and the bound projected at right angles as exactly as could be to the wharf, where a flag was placed at each end, and also a stop-watch. The boat was ordered under way at dead water, or when the tide was found to be without movement. As the boat passed one flag it was struck, and at the same instant the watches were set off. When the boat reached the other flag it was also struck, and the watches instantly stopped. Every precaution was taken before witnesses, the time was shown to all, and the experiment declared to be fairly made. The boat was proved to go at the rate of *eight miles an hour*, or one mile in seven minutes and a half, upon which the shares were signed over with great satisfaction by the rest of the company. It afterwards made *eighty miles* in one day.*

This was in October, 1788, and the boat was called the "*Perseverance.*" On the 12th of the same month, she ascended the Delaware to Burlington, with thirty passengers, a distance of *twenty* miles, in *three hours and twenty minutes*. She had as yet no cabins, but ran as a passenger boat for some time on the Delaware. If Dr. Thornton means by "a day" twelve hours, running time, her speed was *six and two-thirds* miles per hour. Her trip to Burlington was made at *six and one-third* miles an hour. But Mr. Fitch had

* Lives of Eminent Mechanics, p. 32.

calculated upon a regular rate of eight miles, and was therefore not satisfied with her performances. The success already obtained gave him pleasure, but he looked forward to greater results. He was well aware that along the level roads of the Delaware, where stages could make five or six miles an hour, a passenger boat of six miles could never be profitable. It was necessary to exhibit a speed which should astonish the beholder, in order to induce the public to travel upon a craft that had more the appearance of an infernal machine than of a quiet, comfortable, and safe conveyance. He therefore abandoned the company to make new and more enlarged efforts, and the Perseverance was laid up for the winter.

Although no money could be obtained from Congress, the States, or from various corporations addressed on the subject, the legislatures of New York, New Jersey, Pennsylvania, Delaware, Maryland, and Virginia, had granted him valuable but indefinite monopolies for a term of years.

His dissatisfaction with the company arose from repeated interference on the part of the members, upon which he was dependent for money, in the construction of the machinery. In the winter of 1788–9, old scores were settled, and in the coming spring new arrangements were made for a more perfect and enlarged boat. She was completed in the fall, and made an experiment which was satisfactory. That night she took fire and burned to the water's edge.

In the spring of 1790, the second boat was repaired, a consolidated company formed, and a third steamer constructed during the summer. These were destined to fulfil the conditions of the grant from Virginia, requiring the presence of two working boats within her waters before the expiration of four years. The law expired November 7th, 1790, and the two boats were lying in the Delaware ready for a movement, when a furious storm arose and drove one of them upon Petty's Island, where it lay until after the statute expired. Deprived of every hope of relief from the grants of the States, he turned to the new Congress and the patent laws under the constitution. On application to the Congress of the Confederacy in. 1788, he used the following expressions: "We have overcome every difficulty which can cause doubts to arise, having done what was never done before. We have exhibited to the world a vessel going against strong winds and tides, the vessel carrying the engine, the engine propelling the vessel, and all moving together against the current. If we never carry it to any greater degree of perfection, we have, I presume, merited a generous reward."

In the petition of July 2d, 1790, from which we have already quoted, he recapitulates the history of his enterprise, and observes that, "Having at length fully succeeded in his scheme, he trusts he now comes forward not as an imaginary projector."

The year 1791 was principally wasted in securing a patent from the United States, which bear date August 26th of that year. This patent is not to be found in the Commissioner's office, having been destroyed by fire with the public buildings in 1836. Its terms, therefore, can not be quoted. But in 1817, a committee of the legislature of New York, before whom it was shown, with drawings, models, and testimony, make the following official statement respecting it: "In Fitch's boat, the cranks of the axle beam were connected with a frame, from which paddles were suspended, acting in an elliptical line upon the water; while in Fulton's boat, the axle was attached to vertical wheels, with paddles or buckets permanently fixed in the periphery, and in both the motion of the axis was rotary. The boats built by Livingston and Fulton were, *in substance*, the invention patented to John Fitch in 1791, and that Fitch, during the time of his patent, had the exclusive right to use the same in the United States."*

This committee had before them written statements of Dr. Rittenhouse, Andrew Ellicott, John Ewing, and Oliver Evans, who certified to the performances of the Perseverance and other boats on the Delaware. General Bloomfield appeared in person, and stated that he had passed up and down the Delaware as a passenger on the boat, and regarded her as in successful operation.

The patent system was then so loosely executed, that he regarded his parchment from the United States as calculated rather to involve him in useless litigation than to give him valuable protection. He had been in communication with the Spanish minister at Philadelphia, and the Governor at New Orleans, respecting a right to the use of the Mississippi.

Mr. Vail, the United States Consul at L'Orient in France, was desirous to propose the matter to the National Assembly. Although the permission relative to the Mississippi was at length granted, Mr. Fitch concluded to visit France, and did so in the year 1793. But the civil disorders of that kingdom prevented the accomplishment of his desires. He withdrew to London, and by working his passage as a sailor regained America, broken down in body and dejected in mind, and disgusted at the stupidity of a generation who could not, like him, comprehend the immeasurable benefits of the application of steam to water-craft. In 1796, he withdrew to his lands near Bardstown, Kentucky, which he found in the occupation of others. In 1798 he took a fatal poison, and died in the cham-

* New York Review, Vol. iv. No. 7, p. 148.

ber of a tavern, attended by no relative or friend (his landlord excepted), and was buried in the corner of the grave yard, in presence of six or seven persons, without a stone or other monument to mark the spot.

The fulfilment of our design, to do justice to this child of misfortune and forgetfulness, would require an examination of the claims of subsequent inventors to works evidently the property of Fitch. Fulton had an eulogist, who, while the earth was still fresh upon his grave, sounded his praises to the world. They are known wherever a steamboat cleaves the water, on every shore which has echoed with the sound of its engine. But Fitch had none. It was even doubtful, until within about two years, whether his grave could be identified. His manuscripts were sealed up in the year 1792, with directions not to be broken until thirty years after his death. He had withdrawn from the scene of his sorrows and his triumphs to an obscure village of the remote West. He was in his grave, and his relatives were ignorant of the time, place, and manner of his death.

The first model of his boat at Warminster was destroyed or lost. The patent was buried in the archives of the office, until it was consumed by fire. His contemporaries and friends, though retaining a cherished remembrance of his genius and integrity, were scattered throughout the land, advanced in life, or, already like him, in the grave. In 1805, a model, made by him in Bardstown, was burnt in the house where he died. In 1842, the pamphlets and papers left to one of his executors, the Honorable John Rowan, were likewise consumed by fire. Oblivion in every form seems to have settled upon his memory in a cloud of thick darkness, which we hope is about to be dispersed.

In 1828, the Honorable Robert Wickliffe, then a

member of the Senate of Kentucky, who had seen Mr. Fitch while at Bardstown, and knew the important services he had rendered to mankind, proposed a resolution authorizing a plain monument to be erected over his mortal remains. It was referred, on the 8th of February, 1828, to a select committee, who made no report, and the project was allowed to remain without further notice until the winter of 1843–4. At the time of Mr. Wickliffe's resolution the manuscripts at Philadelphia were not opened. Little or nothing material was known of the history of the man whose name it was proposed to distinguish. It is not strange that no interest was excited, no action had, and that the shades of forgetfulness were suffered to gather again about his memory. At the last session of the legislature of the State of his adoption, a memorial, embracing an abstract of his inventions, successes, eccentricities, and trials, was presented at the last moment of their sitting, reviving the subject of a monumental notice. This will be found on the printed Journals of the House for 1844, p. 588, having received no action for want of time. With Mr. Fitch it was a favorite desire that his bones might be laid on the banks of the Ohio. In a moment of despondency, while pursuing his gigantic undertaking at Philadelphia, he exclaims: "Why these earnest solicitations and excruciating anxieties? Why not leave them, and retire to rest under the shady elms on the fair banks of Ohio, and there eat my coarse but sweet bread of industry and content, and, when I have done, to have my body laid in the soft, warm, and loamy soil of the banks, my name inscribed on a neighboring poplar, that future generations, when traversing the mighty waters of the West, may find my grassy turf." And still later he breaks forth in the same poetical strain, referring to the position of his grave, and hoping that it may be made on the shores of some of the waters of the West, in order that the "song of the boatman might enliven the stillness of his resting place, and the music of the steam engine sooth his troubled spirit."

In case the efforts to secure monumental honors for his remains by the act of public bodies shall fail, there is a band of surviving friends who are prepared to do, and have resolved that it shall be done.

The rubbish of forty-five years has almost obliterated his grave, and the tide of time has carried the principal part of his contemporaries into the same eternity with himself. But a few still live, and with them the sense of his merits, the remembrance of his achievements, and the bitter sorrows that were his rewards, are not obliterated or even dimmed by age. It is expected that, before many years have passed, justice, though tardy in its arrival, will be at last rendered, and his genius will receive the homage, as his misfortunes will the commiseration, of his countrymen; that those who navigate the noble rivers of the West will, from the decks of numberless boats, yet witness upon some commanding headland of the Ohio a neat, white, and conspicuous column which shall mark the repose of Fitch.

Delay, instead of operating as a bar to the performance of such a duty, instead of precluding by limitation the obligations of justice, gives new force to those obligations. Other men have been rewarded for similar inventions by the enjoyment of universal celebrity, Fitch by universal neglect. In the Congress of the United States and in public orations, in State legislatures, and even in foreign countries, how often and how honorably has the name of Fulton been pronounced! His is but one of a crowd of inventors whose names rush upon the memory when the steam engine or the steamboat engages our thoughts. There are those of Savary and Newcomen, of Watt and Bolton, the elder and the younger Stevens, of Rumsey and Franklin, Stanhope and Livingston,—these are familiar; but where is the name of Fitch? While the friends of Rumsey, Evans, and Fulton have offered encomiums upon those gifted mechanics, those public benefactors,—while the legislature of the nation has been moved in their behalf as meritorious citizens and inventors, and the treasury of the nation has responded her gratitude in a substantial form,—where is the orator, the voice, or the petition, that has presented the claims of Fitch?

The reality of these claims will appear in a stronger light by comparing more closely the nature of his machine and that of his successors. Practical minds will at once enquire why his boat was not brought into immediate use. This will be made at first view the test, the sole test, of the value of his invention. We are now tolerably well informed of *what that invention was;* we have a knowledge of what in it was *borrowed,* and what was *originated.* I have endeavored to present fairly *what had been done* up to April, 1785. From every thing which was invented by him at that time, Bernouilli, Rumsey, and Franklin, were as far as possible; Raynal and Jouffroy were almost as far. Before Watt the engine was but poorly adapted to *any* machinery: Watt had never adapted it to a boat. Twenty years after, Watt and Fulton undertook the work, and found it a most difficult and doubtful application.

This *adaptation* was effected by Fitch in 1788. A quarter of a century afterwards, the workshops of America could not furnish an engine of any kind. Fitch made one, and put it it operation in 1787–8.

In 1807 a steambat was run from New York to

Albany, at the rate of *four and seven-tenths* miles an hour, by Robert Fulton, and returned at the rate of *five miles* an hour. In 1788 a steamboat ran upon the Delaware *eighty miles* in a "day," and *twenty* miles in *three hours and ten minutes*, which was built by John Fitch. Both had an engine and crank motion, and both had shafts or axles. At the end of the shafts of one were paddles, at the end of the other wheels. Both the wheels and the paddles projected in an equally cumbrous manner over the sides of the boat. The *Perseverance*, with thirty passengers, made *six and one-third* miles an hour; the *Clermont* made *four and seven-tenths*. How circumstances and times alter cases! The trade and transportation of the Delaware in 1788 were limited, and the roads along her banks were level. The *Perseverance* and her consorts paid no dividends.

The trade and travel of the Hudson in 1807 were important, and the journey along her shores rough, tedious, and forbidding. The *Clermont*, at a speed almost one-third less, procured freight and passengers, and by the test of profit and loss, has been declared successful, more successful than her predecessor upon the Delaware. Had they started together, over the same course, at the same time, the *Perseverance*, gaining one mile and six-tenths an hour, would have reached Albany fifty-two miles in advance. Her performances on the Delaware had been equal to that nineteen years before, and on trial in the harbor had shown a speed of three miles greater than the *Clermont*. But the *Perseverance* is forgotten, and the day of her triumph also, while mankind everywhere within the scope of civilization remember the month of August, 1807. The North River boats continued to run, pay dividends, and improve during the life of the monopoly of Livingston and Fulton; but at its expiration their steamers had but just acquired the speed of eight miles an hour. Stevens put them up at once to thirteen or thirteen and a half, being a gain of five miles an hour, or a greater increase than the actual speed of the first class of boats. In the contemplation of Fitch something equal to, or exceeding eight miles, was necessary to success. In practice upon the Hudson, five miles secured an immortal name. Was it for the invention? Mr. Fulton, a modest and just man, never claimed it. Was it for the engine? This was the work of Watt. Was it for the crank, the shaft, or the axle? The New York committee say they were all found in Mr. Fitch's Patent of 1791, and Noah Webster, Esq., says that in 1793 Mr. Vail presented all of Mr. Fitch's papers to Mr. Fulton for examination.* Was it for the wheels? They had been seen and attached to boats more than half a century, and had been tried repeatedly within a quarter of a century.† Was it for speed? In this important particular there was a retrograde of one-third. For convenience or strength? These were subsequent improvements. For perceiving in advance of his day

* Graham's Magazine, July, 1843, p. 108.

† The Philadelphia inventor had tried them, and found that a given power acting upon paddles produced more velocity. The paddles fulfilled the conditions now obtained by an enlargement of the wheel. It saved the expenditure of power, which was lost by forcing the buckets flat-wise into the water, and lifting them from it. It is remarkable that the angle of Fitch's paddles with the water, as they were dropped into it, correspond very nearly with that of the buckets of a thirty foot wheel. The principles of his conclusions upon this angle are thus verified to be more correct than those of the early wheel-boat builders. His results and the results of the best constructed modern boats were the same,—they produced the greatest velocity with the least power. The difference in construction is not a change of the principle. The wheel is more firm, and less liable therefore to injury, but the force of the resisting surface is the same, applied in all small wheels at a less advantageous angle. The first boats of the Hudson had wheels of twelve to sixteen feet in diameter.

the advantages to commerce, and especially the commerce of western rivers? This subject had been presented to the public in every form, to an incredulous Congress, to city authorities, and State legislatures, by Mr. Fitch, during ten long years of importunity and disregard. Was it for predicting a future navigation of the sea by "packets and armed vessels?" Fitch had recorded that prophecy in his letter to Franklin, October 12, 1785,* and to Rittenhouse in 1792.†

As an inventor Mr. Fulton's genius shone most conspicuously, not in ordinary steamboats, but in submarine and marine warfare,—a display measurably forgotten by his countrymen. It was upon torpedoes and submerged batteries, upon steam frigates and cable cutters, that his mechanical talents and his science exerted itself with most splendid effect. Circumstances beyond the control of man rendered the other branch of his study most conspicuous. But why did not Fitch and his companions occupy the Hudson? He was not compelled by the New York statute of 1788 to put boats in motion within that State, as he was in Virginia. His efforts and hopes were first directed to the latter State, and we have seen how sadly they were frustrated by the loss of one of two boats, required by law, at the last moment, when they could not be replaced. We have seen how, at this hour of trial, his resources, his health, encouragement, and future prospects, all failed at once. For all valuable purposes this was the death of Fitch. His prolonged existence was of no further benefit to his invention, yet it was a continuous scene of grief so intense that it can not be recited without exciting sympathy in every breast.

* Life and Writings of Franklin, Vol. x. † Lives of Eminent Mechanics, p. 33.

EARLY DISCOVERIES IN THE MISSISSIPPI VALLEY.

[Cincinnati Directory, 1844.]

In the historical details of the separate cities of the West it would be necessary, in case we should ascend to its remotest occupation, to repeat many events, some of them many times. For instance, a perfect history of the city of Pittsburgh would embrace the annals of the early French discoverers; New Orleans is in the same condition, having been first occupied by the same people in their early enterprises in North America; Natchez and St. Louis, Vincennes and Detroit, have the same origin. From all these places we shall procure a compact yet substantially complete record of their annals and their advance, through the lapse of many generations, to the rank of important commercial cities.

To avoid as much as possible the necessity of a tedious repetition, we offer a condensed view of these discoveries, as they are said to have been made, in the Mississippi Valley. This rapid sketch will embrace all the region now called by the name of the *West*, through which trade has established its channels, and will therefore answer for reference in all cases where the reader is desirous to trace the history of the city of his residence back to the first appearance of civilized man upon its shores.

THE SPANISH ADVENTURERS.

The Spanish nation, among other recollections of the days of their pride and power, refer to the discovery of the Mississippi as the result of their enterprise and courage. They relate the fact that *Ferdinand De Soto*, a companion of Pizarro, and emulous of his fame,—a man who had been Governor of Cuba, —landed at *Espirito Santo*, in Florida, in May, 1539, with nine hundred men and three hundred horses.

With a part of this force he penetrated the forests so far as to strike the Mississippi at the lower *Chickasaw Bluffs*, in 1541. They cross the river, and journey up along its western shore in search of wealthy cities and rich mines of gold, probably as far as New Madrid. Parties of Spaniards went onwards towards the Missouri, but found nothing of the wealth of which they were in search. The little army struck into the country on the west, and finally rested for the winter on the banks of the Washita. In 1542 they returned to the river Mississippi, at the mouth of Red river, called the country of the *Guachoya*. On the 21st of May De Soto died, and his body, as some authors say, wrapped in a cloak, and others enclosed in an oak log, was sunk in the waters of the Mississippi. The Spaniards liberated their slaves, and in the next spring had prepared barges for descending the river.

A large number of these warlike adventurers perished miserably in combats with the Indian tribes, by starvation, and by the fatality of the climate; but enough of them survived to relate their travels, and to furnish their countrymen with a description of the Lower Mississippi.

Whether this relation was known in France at the

moment when her people took up the cause of western discovery, we can not at this remote day satisfactorily determine. Her foremost and most enthusiastic travellers in the new world leave us to infer that their *first* knowledge of the existence of that river was derived from the natives who inhabited the Upper Lakes. But it is a rational supposition that they may have *heard* of the great river of the Spaniards, and yet regarded the stream which the Indians of Lake Huron marked out in their rude style upon the ground as *another* river, occupying a more westerly position, of greater length, and having a different connection with the ocean. The French therefore *claim* the honor of discovering the Mississippi.

THE FRENCH DISCOVERERS.

This people approached the Gulf of Mexico by way of the St. Lawrence and the lakes. In the fall of the year 1535, in the reign of Francis I, a Frenchman by the name of *Cartier* ascended the river St. Lawrence as far as the island of Montreal. Six years afterwards, a fort or trading post was established at Quebec, destined to abandonment in a short time. The site of the old fort was re-visited by Champlain in 1603, and in 1608 an association of merchants of *Rouen*, *St. Malo*, and *La Rochelle*, commenced the city which has since risen to such commercial importance. In 1620 it was well established, and a good fort erected there.

During the year 1625 the Jesuit missionaries, who had already appeared on the banks of the St. Lawrence, carried the cross to the shores of Lake Huron, and in 1634 Daniel and Brebeauf established a permanent missionary station upon its shores.

In 1641 two missionaries, *Ramboult* and *Jouges*,

arrived at the Falls of St. Mary's, and found there two hundred *Nadowessies*, or *Sioux*, disposed to receive the gospel.

From thence, these devoted ministers of the cross pressed, in considerable numbers, into the Indian country; offering the rites of the Catholic religion, baptism and absolution, to all such as would receive them.

The French traders travelled in company with the Holy Fathers, gathering rich cargoes of northern furs from the native tribes along the lakes.

The Abbe *Mesnard* preached repentance to the Sioux, at Kewena Bay, on Lake Superior, in 1660; but, desirous to spread his faith still farther along these inhospitable shores, he departed for Chegoimegon, and perished in the woods near Portage Lake.

Five years afterwards, Father *Allouez* reached Chegoimegon, and erected a chapel there. The Chippewas, Hurons (or Wyandots), the Sioux,—the Illinois and the Ottawas, inseparable allies of the Hurons,—were at this time (October, 1665) in grand council at this place. They spoke of a great river, which they called *Messipi.*

Nicholas Perrot, a layman, and authorized by the French Intendant of Canada, assembled the nations at St. Mary's, in 1671. After much friendly intercourse at St. Mary's, Perrot, Marquette, and Dablon proceed to explore the western shore of Lake Michigan; and Perrot succeeds in reaching the present site of the city of Chicago. *Joliet*, a French officer, was commissioned to find the "great river;" and in 1763, being accompanied by Father Marquette, two Indian guides, and five voyageurs, they ascended the Fox river of Green Bay. On the tenth of June, they drew their canoes across the Winnebago portage, and launched them upon the current of the "Ouisconsin."

The aborigines depicted this stream as dangerous in the extreme; not only by reason of its quicksands, whirlpools, and rapids, but there, and in the great river itself, dreadful demons had taken up their abode, who caught up all passengers in their horrid embrace. The adventurers persisted, however, in their voyage, and, in seven days, floated out into the broad channel of the Mississippi, unhurt. Here, having accomplished the glorious object of their mission in safety, they offer thanks to Almighty God for his protection, standing on the banks of the mightiest river of the world. On its shores, and especially along the borders of the Wiskonsin, they beheld such scenery as no traveller had seen before them—the rolling upland prairie, spread out beyond the sweep of vision, on every side a meadow clothed in the deepest green. They saw every variety of game feeding on the hill sides, which was easily brought down by their muskets; and having feasted their bodies upon the fish, fowl, and wild meat of the region, and their eyes with the surpassing beauty of the great valley the threshhold of which only was passed, they committed themselves to the guidance of the newly found river of their desires.

They floated onward to the mouth of the Ohio which they call the "*Ouabache*," where they found a band of the Shawnees residing. They even continued to sail downward, to the mouth of the Arkansas; bu here the expedition terminated.

Joliet returned to Quebec, by way of the Illinoi river and Chicago creek, to lay the results of th journey before his patron, the Intendant Talon. Talon was overjoyed to learn that his Nouvelle France in addition to lakes of greater extent and purer wate than any in the known world, embraced a luxurian domain, equal to all Europe, and rivers upon a scal of greater magnificence than the other displays of nature in this hidden world.

Marquette remained with the Miamies about the south end of Lake Michigan until May, 1675, inculcating the precepts of the Gospel. He was already advanced in life, and exhausted by travel and exposure. Coasting along the eastern shore of the lake, he steered his canoe into the mouth of a creek which now bears his name. He landed upon the shore and retired amid the trees, fragrant with the buds and flowers of spring, to pray in secret, knelt down upon the ground, and was soon after found dead at the same place.

But mankind have awarded to the Chevalier *La Salle* the merit of making the first thorough *exploration* of the Mississippi, and of establishing settlements upon its waters. He constructed the first vessel which spread her white wings upon the waters of Lake Erie. The *Griffin*, a sail-craft of sixty tons, was launched at the mouth of Tonewanda Creek, on the 7th of August, 1679. She pursued her course through the lake, the Detroit river, and the Lake of the Hurons, to Mackinaw, on the peninsula of Michigan, where a trading post and fort were established.

The Griffin was loaded with furs on her return, to the great joy of Monsieur La Salle. In her were the results of many years of incredible exertion, his entire fortune, and with it the resources that were to sustain his enterprise. She foundered on Lake Michigan. The adventurer took canoes, and coasted down the eastern shore of Lake Michigan. He ascended the St. Joseph's, and finding a low swampy tract which communicated with a southern stream, he transported their light vessels into the *Kankakee*, and down it to the Illinois. On its banks, they built the fort of *Crevecœur*, or the "broken heart;" signifying the forlorn state of their feelings at the time. Here they winter in 1679–80, and await the return of the Griffin.

It was not until 1682 that La Salle himself descended the river, and determined to find its discharge into the ocean.

At the Chickasaw Bluffs he erected a cabin, at the mouth of the Arkansas raised a cross, and, sailing with the principal channel, on the 9th of April he saw the Gulf of Mexico. Here he planted the arms of France in token of possession, and returned by way of Crevecœur and the lakes to the city of Paris.

To the French court the affair appeared to be of high importance. It was determined that the "*Meschasebe*" should become the residence of Frenchmen, and La Salle was provided with *four vessels*, one hundred soldiers, and one hundred and eighty artisans, to effect this object.

They reach the Gulf of Mexico, but pass the mouths of the Great River without being able to find them, and landing finally at the Bay of Matagorda, erect Fort St. Louis, in June, 1685. After suffering at this post a year and a half, the Chevalier departs by land in search of the "fatal river," with a company of sixteen men. Twenty-four, the remnant of his armament, remain at the Fort. On the Trinity River one of his men kills his nephew, and when La Salle inquires after the young man, they strike him to the earth, where he dies.

But the system of colonization was not abandoned on account of the loss of its pioneer. La Salle had established a fort and a trading post on the St. Joseph's of Lake Michigan, called Fort Miami; and his party, soon after his arrival on the Illinois, located a station on the Kaskaskias. Monsieur D'Iberville reached the mouth of the Mississippi in 1699, as the successor of La Salle, built a fort, and ascended to the village of the Natchez. As he crossed the low ground after landing, and walked up the hill in view of the Indian town, he was so much delighted with the beauty of the spot that he immediately traced out a fort, and called it *Rosalie*, after the Duchess of Pontchartrain.

In returning he met an English ship of twelve guns, claiming for the crown of Great Britain the entire region which the French had traversed. They (the English) insisted that Cabot had discovered the entire coast of Florida, as this region was then called. But being at once ordered to depart, they did so, threatening, as they went, to return in the following year with force, and drive away these French interlopers. But they did not come, and the next year more Frenchmen arrived, who made the first settlement at the Isle of Dauphin. In 1712 this place was abandoned, but "Maubile" and Biloxi continued. This was the year of *Crozat's* grant, bounded on the north by the Illinois.

Crozart resigned his monopoly in 1717, and *John Law's* great "Company of the Indies" take possession of Louisiana, as Monsieur D'Iberville chose to call it. The Governor, M. Bienville, selected New Orleans as a post of trade and defence. In 1719, France and Spain being at open war, the French fall upon Pensacola, a Spanish town, which they capture and occupy; but the Spaniards repossess themselves of the place, and the French colonial troops again invest and carry the fort. Hitherto the Council General of the colony had been fixed at Biloxi, but in 1721 the officials and their attaches removed to New Orleans, a place which, from that hour, has not ceased to be a capital.

New France was therefore divided into two provinces,—Canada and Louisiana; but what constituted New France, geographically considered, was by no means reduced to a certainty. By the treaty of Utrecht, in 1713, the French king had yielded Newfoundland and Hudson's Bay to the English. The latter colony began at a point or promontory of the Atlantic, in latitude 58° 30′ north, thence to Lake *Mastassin*, thence south-west to the 49th parallel, and with it west indefinitely. New York claimed all north of the 40th degree, and west of New England and the Ottawa river, even to the 49th parallel, Virginia all between 36½° and 40°.

The French had planted a colony at Port Royal, on the Carolina coast, in 1652, which, being abandoned, was revived in 1654, as a refuge for the persecuted Huguenots or Calvinists. But Melendez D'Aviles of Spain, armed with a commission to destroy all heretics, fell upon Fort Carolina and took it, September 25, 1655.

Because Melendez had massacred all the Protestants taken at Fort Carolina, *Dominic De Gourges*, a Calvinist of Gascony, determined upon revenge, and providing himself with three ships and one hundred and fifty men, he sailed to Florida, and surprised three Spanish ships at the mouth of the St. Matheo, by us called St. Johns. He repaired thence to Fort Carolina, and took possession of the post. Melendez having massacred the Huguenots, as he said, not as Frenchmen, but as Calvinists, De Gourges hung his prisoners upon a tree, and attached to it a notice that they were not executed as Spaniards or mariners, but as pirates, robbers, and murderers. But France formally relinquished the whole South Atlantic coast, and Spain resumed it as far north as Cape Fear.

In 1748 the French insisted upon the following line as dividing them from the British colonies:—Beginning at the mouth of the Apalachicola river and the Gulf of Mexico, thence up the same to its source, and with the Allegheny ridge to the sources of the Susquehannah, thence in a straight line to Fort Cohasser on the Connecticut river, near Long Falls, and from this point north-eastward, parallel to the St. Lawrence, to the Kennebec, with it to the sea, and across the Bay of Fundy to Cape Canso.

The English offered to accept of a boundary,—for the north the lakes and St. Lawrence, and the west a line from Presque Isle on Lake Erie, through French Creek to the Apalachian range, as claimed by the French themselves. But the treaty of Aix-la-Chapelle was signed, and by it nothing was settled respecting their colonial limits.

Thus the war of 1754 came on without any fixed understanding of boundary; in fact, it occurred principally in consequence of the common title set up by the two nations to the Ohio country. The "Ohio Company" was authorized by the British Parliament, in 1749, to locate 600,000 acres on the Ohio and its waters; and to have an exclusive trade with the Indians.

Christopher Gist, their principal agent, with his surveyors, entered the country in 1751, and explored the Great Miami. In 1752 he established a trading post and temporary defences on this river, at the mouth of Loramie's Creek, of which the French soon had information, and came with an armed force to capture the station. This they accomplished very easily, took the English prisoners, killed fourteen Piankeshaws who sustained them, and carried the goods collected there to their forts on the Miami of the lakes. And to prevent the Ohio Company realizing their expectations, they took possession of a new route, covering it with posts. From Presque Isle they made a portage to French Creek, and erected a fort upon it. At its mouth, on the Allegheny, they construct Fort Venango, and provided it with a garrison. The Governor of Virginia, regarding those posts as clearly within his colony, considered the proceeding as nothing less than an invasion, and, to to enquire into the matter, sent George Washington, in the fall of 1753, with a letter to St. Pierre, the commandant at Fort *Du Beuf*, on French Creek.

In the spring Governor Dinwiddie raised a few men, and ordered them to construct a fort at the mouths of the Monongahela and the Allegheny rivers. Ensign *Ward* and forty soldiers had scarcely begun to cut pickets when Monsieur *Contreceur* descended the Allegheny, accompanied by a formidable body of Indian and French troops. They took possession of the Virginians as prisoners, and built Fort Du Quesne. The war may now be said to have commenced. It resulted in the treaty of Paris, in 1763, by which France ceded to England all her claims and territories in North America east of the Mississippi to the river Iberville and Lake Pontchartrain.

At the beginning of the war the French had in possession thirty-eight garrisons and trading posts, located as follows: —

Beginning at the north, at the head of the Bay of Fundy, Fort *Chignecto*.

A fort at the head of *Bay Verte*, on the opposite side of the Peninsula of New Brunswick.

Fort *St. John*, mouth of St. John's river.

Cohasset, on the Connecticut, below Long Falls.

Fortifications around Quebec.

Fort Sorel, on the St. Lawrence, at the mouth of Sorel River, west side.

Fort Chambli, on Lake Champlain, at the head of Sorel river.

Fort Frederick, or *Crown Point*.

Frontenac, near Kingston, and a fort at the portage between the Ottawa river and Lake Abbittibis, near Lake Simcoe.

At Niagara, mouth of Niagara river, east side.

Fort Erie, opposite Buffalo; a fort at Presqûe Isle, now Erie, Pa.; another at the end of the portage to French Creek, called Le Beuf; Fort Venango, at its mouth on the Allegheny, then called the Ohio.

Fort Du Quesne, at the forks of the Ohio.

Fort Sandusky, near Sandusky city.

Fort of the Miamihas (Miamies) on the Maumee, not far from Defiance; Fort Pontchartrain, at Detroit; St. Joseph's, on the river St. Joseph's, of Lake Michigan, some miles from the Lake; one at Mackinaw, on the main-land, south of the island; one at St. Francis Xavier, a short distance above Green Bay, on the Fox River; Fort La Roche, and fort of the Miamies on the Illinois, near each other on opposite sides of the river, above Lake Peoria; Fort Orleans, on an island in the "Missuri," above Grand river.

Fort St. Louis, nearly opposite Cahokia.

Fort Chartres, a permanent work near Kaskaskias.

A stockade at the mouth of the Wabash and mouth of the Ohio.

Fort Massac, on the Ohio, and Fort Vincents, on the Wabash; the fort of the "Ouatanons," at the portage from the Wabash to the Miami of the lakes, above Eel river. Some authorities place a post at the mouth of the Scioto, on the Kentucky side. There was a station near St. Mary's, in the county of Mercer, Ohio, at the portage between the St. Mary's and the Great Miami. Fort Kappa, at the mouth of the St. Francis, Mississippi (southern bank), at the mouth of the Arkansas, on the north bank, and stations up the river. Fort Rosalie at Natchez, Fort Balise at the mouth of the Mississippi, a fort on Isle Dauphin, Fort Canada at the head of Mobile Bay, and Fort Toulouse on the Alabama, latitude 32° 20′ north.

An English author, writing in 1747, says of these works, that many of them are mere "extempore stockades," which the French, "by way of ostentation," call forts, and they "are a great nuisance to our American colonies." Of their troops, he observes, there are twenty-eight marine companies, composed principally of "racaille or gaol-birds" from France, who can not be depended upon."

In 1760 the English were in possession of Oswego, a post which had been a French establishment.

We have thus given a cursory sketch of the discoveries and occupation of the French in the Valley of the Mississippi to 1763. The English, like the Spaniards and French, have pretensions to the discovery of the same region.

EARLY ENGLISH DISCOVERIES.

The British, whose vessels sailed into the Mississippi in 1698 and 1699, based their right upon the discovery of *John Cabot*, and Sebastian his son, who saw and explored the Atlantic coast of the northern States, in 1497. The son spent twenty years in these explorations, but always at the north, and is not known to have entered the Gulf of Mexico. In 1583, Secretary Woolsingham dispatched vessels of discovery that entered the river St. Lawrence, and *Douglass* insists that their flag was seen on its shores as early as 1527. Captain Thomas Hutchins, of the 60th regiment of foot, in his account of Florida and Louisiana, asserts that Colonel Woods traversed the mountains from Virginia in 1654, and reached the Mississippi; and also, that Captain Bolt performed the same journey in 1670.

Colonel Spotswood crossed the Alleghenies in 1710, for the purpose of establishing a land company in the colony of Virginia, of which he was Governor. A Virginian, Doctor Thomas Walker, passed the Cumberland Gap in 1750, who was followed by English traders descending the Ohio, during the next year. This was the year of the occupation and surveys of the London "Ohio Company." In Coxe's "Collection of Voyages" (A. D. 1741), there is a map of the "*Sakagoula*," or "Mescha" (great) "Cebe" river of the West; and in "Douglass' Summary," published in 1760, Mr. *Huskes* has placed a *map* of *the West*, embracing not only the Mississippi and Missouri, but a large river heading near the latter, and leading with a navigable current to the Western Ocean. In 1752, Lewis Evans published a map of Kentucky; but none of those geographers appear to have explored the regions portrayed upon their plans. The details of the visits of Woods, Bolt, and Walker, were wanting; and, consequently, much doubt is thrown upon their statements.

But Gist and Washington kept regular journals of their travels, which are still preserved. The Iroquois, who were in alliance with the colonies of New York and Virginia, effectually prevented the French from passing from Lake Erie to the head waters of the Ohio, until 1739, when Monsieur *De Longeuil* reached the Allegheny, and descended the river in a pirogue.

Gist's survey of the Ohio, and of the Great Miami to Loramies in 1751, and the establishment of a post there in 1752, are the first substantial acts of English occupation west of the mountains. The second was the arrival of Ensign Ward and forty-one men at the site of Pittsburgh, two years after. Both Loramie and Pittsburgh were immediately captured by the French. The possession of the Ohio country was however, finally secured to the English, by the recapture of Fort Duquesne in 1758, under General Forbes.

As to their rights by discovery on the St. Lawrence, although Charles I. had authorized *David Kerkte*, a protestant refugee from France, to invade Canada; and, although in 1629 he captured Quebec, the principal city, it was restored by treaty in 1632. Yet, in 1711, Queen Anne contended that the French held of her as *fiefs*; and in 1712, a heavy expedition of sixty-eight vessels and six thousand men advanced against the city of Quebec. A tempest destroyed many ships and one thousand men in the St. Lawrence river, and the enterprise was abandoned.

Thus the real basis of their claims to the St. Lawrence, the Lakes, and the Mississippi, is that of conquest in war from the French; an achievement completed by Wolfe in his last victory at Quebec.

We are thus again at that period when so many political changes occurred in North America, the year 1763.

THE ENGLISH AND SPANISH DOMINION.

The treaty of Paris, February 13, 1763, transferred Canada and most of Louisiana to England. England, at the same time, relinquished Havana to Spain. Spain, in return, ceded Florida to England; and, in April, France, by a secret treaty, yielded Louisiana to Spain.

It was more than a year before the French on the Mississippi were informed of this transfer; and five years passed before the arrival of the Spanish Captain-

General, Don Antonio D'Ulloa. The colonists were so much displeased, that his successor, O'Reilly, thought it necessary to have three thousand troops at New Orleans, to hang six of the principal citizens, and shoot five of the crown officers. There were, besides the Captain-General, a civil officer, called an Intendant, who appears to have been a kind of court of appeals from the commandants and vice-governors. A Sub or Deputy Governor resided at St. Louis. The military commandant of each post exercised, in the absence of the Governor, supreme, civil, and military authority. On complaint, he notified the defendant that he must forthwith do justice. This being disregarded, the offender was ordered to appear, and submit to judgment. If he failed to answer that summons, a file of soldiers brought him into the presence of the Commandant, who administered justice according to his own ideas of right, and the laws of Spain. Writers differ as to the equity with which this system operated on the inhabitants. Mr. Breckenbridge asserts, that the system was mild, just, and acceptable to the people. Mr. Flint remarks, that the Commandant was, in general, an ignorant and despotic man, whose legislation and execution centered in his cane; that the government may be summed up in a few words, viz: a commandant, a priest, a file of soldiers, and a "*calaboosa*."

On the left bank of the Mississippi and the Lakes, the colonial institutions of Great Britain were in operation, so far as inhabitants were found, over whom their sway could be exercised. The French, at their villages, forts, and trading posts, generally retired to Canada or the Spanish towns. A few still prosecuted their trade among the Indian tribes. In 1760, two hundred persons had collected about Fort Pitt. On the Monongahela, a settlement had been formed prior to 1758, called the Decker Settlement, which being cut off by the Indians, restrained the frontier adventurers until 1765. In 1773, a company of discharged soldiers from the Virginia militia descended the Ohio to the Falls, and located their bounty lands.

A grant of Indiana had been made in 1768, to Samuel Wharton, William Treat, and George Morgan. But what constituted Indiana does not distinctly appear. Wheeling had been established about sixteen years, and was becoming a place of consequence in western affairs in 1774.

Colonel George Rogers Clarke, and three hundred soldiers, the troops of Virginia, reached Corn Island, at the Falls of the Ohio, in the spring of 1778, which resulted in a settlement and cultivation of land. The war of the Revolution then raging on the east of the mountains justified Colonel Clarke in making an assault upon the English posts at Kaskaskia and Vincennes, in the succeeding fall and winter. He took both these garrisons, and kept possession of them till the peace of 1783.

The colonial troops had, in the mean time, garrisoned Fort Pitt, Fort M'Intosh, Fort Laurens, and the Fort at Point Pleasant; expeditions had been made against the Indians in alliance with Great Britain; and a virtual conquest of the country, bordering the Ohio on the north, had been made in the name of the Colonies or of the Confederation of the States. This was acknowledged and confirmed by the treaty with England, January 10, 1783, twenty years after she had acquired the country ceded from France. From that moment, one-half of the Valley of the Mississippi, shaking off the monarchial principal, became republican soil. The right bank of that stream retained its feudal and absolute character twenty years longer. In 1789, the Spanish minister, *Count*

D'Aranda, proposed to make three kingdoms in America, one for each of the *infants*. By this scheme, her territories were to cease to be colonies; but monarchy, in its odious form, was to be made hereditary on the banks of the Mississippi.

Another of the Spanish projects in America had been made known the year before; and it was a proposition to the Americans, west of the mountains, to form a separate empire, in consideration of the free navigation of the Mississippi. These designs all failed of execution. At the close of the eighteenth century, Bonaparte began to form projects relative to America, and persuaded Spain to cede Louisiana to France. The Spanish forces had made a conquest of Florida, from England, during the American war, and now held an immense territory in North America. On the first of October, A. D. 1800, a treaty or convention was signed by France and Spain at St. Ildefonso, by which Louisiana passed into the hands of Napoleon. It was confirmed and reiterated in the treaty of Madrid, March 21, 1801; but it was stipulated, that in case of a disposal of the territory by France, Spain should have the refusal of it. France thus became possessed a second time of the fruits of her early discoveries on the Gulf of Mexico.

THE SECOND DOMINION OF FRANCE.

New circumstances arising, gave rise to changes in the policy of the Consul in regard to Louisiana. The assumption of civil authority was delayed. *Laussat* and perfect and *Aymi* the chief judge, at last arrived at New Orleans, in the winter of 1803; but the Captain-General *Victor* was prevented from leaving the Dutch coast with his armament, by the war with England breaking out afresh after the peace of Amiens.

Bonaparte to prevent the English from making a conquest of the territory, and to procure money for his vast civil undertakings, determined at once to sell this province to the United States. Without knowing of his determination, Mr. Jefferson, President of the United States, dispatched a special envoy (Mr. Monroe), to negociate for the Island of Orleans. The population of the American side of the valley had expanded to the number of eight hundred thousand, and the free use of the river became indispensable.

On the 27th of October, 1795, the Spanish government had granted a right of depot for three years; and by mutual consent it had continued without repeal until the 16th of October, 1802, when the Intendant *Morales* suddenly brought it to a termination. There was a party in the country, particularly at the West, who were for taking military possession at once. Mr. Jefferson had no farther design than to secure the remaining portion of the left bank, and instructed Mr. Monroe to offer $2,000,000. A motion was made in the Senate, that $5,000,000 and 50,000 troops be placed at the disposal of the President, to take possession of New Orleans; but it did not carry, and Mr. Jefferson relied upon pacific measures.

Mr. Monroe, on reaching France, was astonished to learn that the French ruler had already decided to sell, not only a part, but all of Louisiana; and the only question to be discussed was the price to be paid for it. On the 10th of March, the Spanish authorities, being still in the exercise of their functions, consented to give us a place of deposit in New Orleans for western produce. But, unknown to all on this side of the water, and to Spain and England on the other, the treaty of cession was maturing, and, on the 30th of April, was settled by the commissioners. The

United States gave 80,000,000 of francs, from which 20,000,000 were deducted for spoliations upon our commerce.

As soon as the papers were signed, the three negociators, *Barbe Marbois*, *Mr. Livingston*, and *Mr. Monroe*, transported with sentiments of joy, that so great a matter had been disposed of to the mutual honor and satisfaction of all, rose and grasped each other's hands with the utmost enthusiasm. Mr. Livingston exclaimed, "We have lived long, but this is the noblest work of our lives. The treaty which we have just signed has not been obtained by art, nor dictated by force. Equally advantageous to the two contracting parties, it will change vast solitudes into flourishing districts. The United States will re-establish the maritime rights of all the world, now usurped by a single nation. The instruments we have just signed will cause no tears to be shed; they prepare ages of happiness for innumerable generations of human creatures."

The article guaranteeing protection to property and the enjoyment of liberty, with the free exercise of religion, was drawn up by the hand of Napoleon.

"Let the Louisianians," said he, "know, that we separate ourselves from them with regret; that we stipulate in their favor every thing which they can desire; and let them hereafter, happy in their independence, recollect that they have been Frenchmen, and that France, in ceding them, has secured for them advantages which they could not obtain from an European power, however paternal it might have been. Let them retain for us sentiments of affection; and may their common origin, descent, language, and customs, perpetuate the friendship."

The Spaniards were now required to execute the treaty of St. Ildefonso. They accordingly delivered the forts and posts on the Mississippi, to Monsieur Laussat and his agents, on the 30th of November, 1803. The reign of France was short and provisional. On the 26th of December, the French prefect, the American governor, Claiborne, and General Wilkinson, commanding the United States' troops, who had entered the city as the Spaniards embarked, assembled at the City Hall. Laussat made a formal transfer of the province, and Claiborne received it in execution of the treaty.

While this ceremony was passing in the Hall, the American flag was brought to the foot of the flag-staff, at the top of which floated the colors of France. As one rose, the other descended, and meeting midway, remained some moments mutually entwined. When the flag of the Union rose in the air, the Americans could no longer suppress their shouts of joy; but the French guard, alive to the scene, expressed the deepest regrets, and as a last homage to the illustrious banner of their country, the leader wrapped it around his body, and paraded the streets at the head of his troops, and finally deposited this symbol of the power and glory of France with the late prefect, Mr. Laussat.

THE DOMINION OF THE UNITED STATES.

A few words will express what we have to say of the power now in the possession of the Mississippi.

Congress divided the territory of Louisiana on the 20th of May, 1804, and the northern territory was attached to Indiana. On the 22d of January, 1812, the State of Louisiana was formed by adopting a republican constitution.

In 1805, Governor Harrison of Indiana, divided Upper Louisiana into six districts or counties. These districts had, in 1810, the following population, of whom 8,011 were slaves:

St. Charles,	3,505
St. Louis,	5,667
St. Genevieve,	4,620
Cape Girardeau,	3,888
New Madrid,	3,313
Arkansas,	1,067
Total,	21,845

Both banks of the river, under the impulse of our people, who derive their enterprise from the happy and free nature of their government, are now occupied by constitutional States, as far north as the latitude of the lakes.

Population has spread itself to its source, and commerce enlivens its entire length. In 1712, there were supposed to be but twenty-eight families resident on the Mississippi and its waters. Now, there are 6,000,000 of souls.

RIGHTS OF AUTHORS OUTSIDE OF COPY-RIGHTS.

[Hesperian, September, 1839.]

We know of no proceedings in State Courts, relative to the rights of authors and inventors, since the Constitution went into operation in 1789. It appears to have been generally believed, that a provision of the 8th section of the 1st article of that paper, withdrew from the respective States all control over the subject; that the grant to Congress of a power "to promote the progress of science and the useful arts, by securing for limited times to authors and inventors the exclusive right to their respective writings and discoveries," vested in that body exclusive authority in those matters; and, through it, in the federal courts, exclusive jurisdiction of cases relating to those rights.

Chief Justice Kent advanced a contrary doctrine, in Livingston *vs.* Van Ingen, 9th *Johnson's Reports*; but we know of no legislative action corresponding with the suggestion put forward in that case. He said, "that if an author or inventor, instead of resorting to the Act of Congress, should apply to the State Legislature for an exclusive right to his production, there is nothing to prevent the State from granting such exclusive privilege, provided it be confined in its exercise to their particular jurisdiction." We propose, however, to view the subject of *literary property* in the United States, in a different aspect, as coming under the judicial, and not the legislative power. If an author, or inventor, in Ohio, or other State, enters the proper court with a complaint that his book, or invention, has been pirated by a person within its jurisdiction, *can such State Court* refuse cognizance of the injury? Our Constitution asserts, that courts of justice shall always be open for redress of injuries to the person, or property, or reputation of any individual; and to disclaim consideration of the case we have just put, a clear exception must be furnished by the court, showing a want of jurisdiction over the subject matter. The excuse should designate another tribunal or authority, having the power and the disposition to entertain the case or the existence of a prohibition by a law superior to the State Constitution.

There is still another ground of refusal, which is, a denial that the work of our author *is property*. If such productions *are property*, and no restrictions are placed upon the States in guarding and protecting it, by the language we have quoted from the eighth section of the first article of the Constitution of the United States, authors and inventors can no more be driven from the seat of justice unheard, than the creditor, or the slandered man. A review of the subject and the discussion will almost necessarily mingle the consideration of a right of property with that of jurisdiction.

The first English statute upon the subject is the 9th of Anne, c. 19, enacted in the year 1709. It provided, that within a certain period (fourteen years) the author "shall have the sole right and liberty of printing, vending, etc.," his work. The penalties of the act were, that every copy should be forfeited, to be *destroyed*, and a penny a sheet paid to the informer who should prosecute. The existence of a common law right, and the effect of this statute upon it, may be gathered from the English cases.

The King's Bench in 1770, decided, "that any author had a common law right in *perpetuity*, to the exclusive printing and publishing his original compositions." 2 *Kent's Com.*, 307; 4 *Burrowes' Rep.*, 2203; *Taylor* vs. *Miller*. Injunctions to prevent the publication of manuscripts without copyrights were frequently granted upon the ground that there was a *property independent* of the statute. *Eden on Injunctions*, 199, 200; 2 *Eden*, 329; 2 *Merivale*, 435; 2 *Atk.*, 342; 2 *Ves. & Bea.*, 19.

In 1774, the House of Lords reversed the decision in Taylor *vs.* Miller, by deciding that the common law right, if any existed, could not be exercised *beyond the time limited* by the statute of Anne. *Donaldson* vs. *Becket*; 7 *Bro. P. C.* 88, and *Burrowes*, 2408. This case turned upon the question, whether the statute abridged or took away the common law right. And to allow this case only the force to which it is entitled, the circumstances attending these leading decisions will be given. In Miller *vs.* Taylor, all the judges, except Justice Yates, subscribed to the existence of a common law right expressed in the decision above quoted, and Lord Mansfield was one of the court. Four years afterwards, in Donaldson *vs.* Becket, the Lords referred the question, how far the statute affected the common law right, if it existed, to the twelve Judges of England. Lord Mansfield declined giving an opinion, but adhered, in sentiment, to the case of Miller *vs.* Taylor. Eight of the remaining eleven agreed, that a common law right existed before the statute. Six were of opinion that the statute abridged or took away that right; and five that it did not, who, with Mansfield, would have divided the Bench. Upon this authority rests the right of "property," under the statute of Anne, from which, as will be shown hereafter, the act of Congress of 1790 was almost copied.

But, twenty-four years after, in Beckford *vs.* Hood, the King's Bench declared (all the judges concurring) that the statute *confirmed*, for the time named in it, the common law right, and that the penalties, instead of barring the common law *remedy*, constituted an *accumulative* right of action. 7 *Term. Rep.*, 623.

At this period, 1798, our statute had been in existence eight years, and may be said, therefore, to be interpreted by the King's Bench, so far as it agrees with the English statute of that day. It is now proposed to give the American statutes in substance, and we will confine ourselves to those parts bearing upon this discussion.

The words corresponding, in the first section of the law of Congress, of April 31, 1790, with those of Anne, are these: "Shall have the sole right of printing, re-printing, publishing and vending," books, maps and charts, for fourteen years, and at the expiration, if the author is living, renewable for fourteen years, in addition.

Section second gives a penalty of fifty cents a sheet, one-half to the lawful author who prosecutes, and one-half to the Government. It also forfeits all the copies, for the *purpose of destruction.*

Section sixth relates to piracy of a *manuscript*, and says, *the author may sue for damages, in any court having jurisdiction*, in an "action on the case." The act specifies no particular jurisdiction, and therefore relates to the Federal Courts generally; and were it not for the division of powers between the States and the National Government, by our Constitution, there would be very little difficulty in adopting, almost entire, the English practice, under the statute of Anne. What, then, was the effect of the Constitution, as adopted in 1789, and the laws enacted in pursuance to it, April, 1790, upon the rights of authors?

Prior to the revolution, British North America was doubtless governed on this point by the English law. Between the Declaration of Independence in 1776, and the Constitution, the States, though freed from the political rule of the mother country, in general, adopted her methods of administering justice and expounding and enforcing private rights. We may safely conclude, that this subject stood, at the adoption of the Constitution, upon the same basis as it did in England. The words of that instrument are intended to *secure*, and not to *create*, a right of property. The cases of Taylor *vs.* Miller, and Donaldson *vs.* Becket, were before the Convention of 1787, and all the law and the arguments connected with those prominent decisions were familiar to the delegation. By them a common law right existed in the author *before* the statute, and the judicial exponent of law in England was equally divided upon the question whether it existed *under* or *during* the statute. *Every decision* recognized the right to a common law *remedy* under the statute; and the necessity of pursuing it in that way was apparent, from an absence of statutory provision respecting the form of action. The act of Anne, then, according to Beckwith *vs.* Hood, *secured more fully* to authors this property in books, by ordaining forfeitures and penalties, in addition to the usual remedy for injuries to their interests.

The acts of Parliament are, in such matters, in England, like our American Constitution here, the supreme law of the land. The phraseology of our law, made in pursuance to the Constitution, is almost identical with the enactment of Parliament. Is there anything in circumstances to give the same terms greater scope on this side the water than they had on the other? If not, where is the remaining right, as yet untouched by the Constitution or the law?

The Federal Government took nothing, in any of its branches, but what the people bestowed in the Constitution. They gave Congress the power, the *ability*, to promote the useful arts and general science. This is the prerogative yielded up to the Union, and it was either concurrent or exclusive. The exercise of it by Congress was not *compulsory*, but *voluntary*, which seems to give it the character of a concurrent power, that the State Legislatures might exercise in the absence of, or in subordination to, the General Congress.

The *manner* and also the extent or mode of the encouragement to be given, was pointed out in the same connection in the eighth section, viz: by *securing* to authors, etc., exclusive rights. Could Congress, under such a wording, say to the people of the States, that, not complying with the terms of our law, you shall have no right or interest in your literary productions? The National Legislature is empowered to secure, not to abrogate a known, existing, and well-defined right. This would prove a great contradiction to the avowed objects of the Constitution, and would obtain its introduction in a very insiduous manner. The case would form an exception to the professed intention of that document, which is the protection, not the destruction of property. Apply the doctrine to a literary work in being at the adoption of the new form of Government in 1789. The author considers the common law right, or the State authority, a sufficient protection for his book, and does not think it necessary to assume the trouble and expense of a copyright. At the moment when the new Government came into existence, according to a construction of this kind, his rights all fail, and the product of his pen is at the mercy of any invader. And in searching for the clause which has produced such an unexpected change in his property, it is discovered to be a phrase for the promotion of learning and art, by *farther securing* for limited times, to the author, the sole use of his book. Assuming then, that Congress can *create* no right of property, but only more effectually protect, for a time, what is in nature and by common law already in existence, the inquiry occurs, where is the remnant of authority and jurisdiction not delegated to the federal power? Can it be any where but in the people of the States? or, if by them delegated to the Local Governments, then in some branch of the State authority?

But to bring forward the proceedings of Congress to a late period, the acts subsequent to 1790 should be here introduced. Upon the 8th of Anne, similar to the first American law, Lord Kenyon had remarked, in Beckwith *vs.* Hood, "Nothing can be more incomplete as a remedy, than these penalties; for, without dwelling upon the incompetency of the sum, the right of action is *not given to the party grieved.*" Our statute differed from this, in giving *one-half* the penalty to the *party grieved*, if prosecuted within *one year.* The expense and trouble of prosecution would consume most of the author's share of the judgment; and conceding this to be the only remedy in the United States, the progress of a year in time, without suit, leaves him without relief. April 28, 1802, another act was passed, relative to historical and other prints, with forfeitures and penalties, like the law of 1790, but in no manner altering or affecting that act. The limitation in this act is two years. The next Legislation was perfected February 3, 1831, forty-one years after the original action of Congress, a period in which the property in books, maps, and charts, in the United States, was apparently subject to similar common law incidents, with the same things in England. The first section extends the subjects of copyright, including with books, maps, and charts, musical compositions, prints, cuts, and engravings.

Section six creates the penalties for printing, publishing, vending, importing, or offering for sale, any of the articles enumerated in the first section, without the consent of the author. The books are here forfeited to the lawful owner, and a fine of fifty cents a sheet is recoverable, one-half to the author or owner who prosecutes. The maps, charts, etc., are forfeited to the same use, and also the plates, with a fine of *one dollar* per sheet or copy, one-half to the owner. Limitations to prosecutions, two years. The changes or amendments here adopted are in favor of the author, by bestowing the surreptitious copies upon him in full, instead of ordering their destruction.

Section nine has the following provision: "If any unauthorized person shall publish or print 'any *manuscript whatever*,' he shall pay all damages, in an *action on the case*, in any court having jurisdiction; and the United States Courts are empowered to grant injunctions against the issue of such publication."

We do not know the judicial construction to this section. The other parts of the act quoted do not differ in principle from the acts of 1790 and 1802, which are nearly equivalent to the *8th Anne, chapter 19*. The language of the section seems to intend to *bestow* a general jurisdiction by an action for damages, and a particular authority in cases of injunction arising out of a wrongful publication or printing of a *manuscript;* and the inference is, that Congress conceived themselves authorized to allow a suit for damages, in the cases of books, maps, and charts, but did not choose to do so.

On the 15th of February, 1819, a law of Congress was passed, giving to the *Circuit Courts* of the United States original cognizance of suits at law and in equity, *arising under the laws* of the United States, respecting writings, inventions, etc. This act is a mere disposal of jurisdiction among the federal courts, creating no fresh *rights of action.* At the same time it had been decided, in Robinson *vs.* Campbell, 3 *Wheaton*, 221, "that, by the laws of the United States, the Circuit Courts have cognizance of all suits of a civil nature, *at common law and in equity*, in cases which *fall within the limits* prescribed by those laws." And *remedies* in the federal courts were declared to be according to the principles of common law and equity, as distinguished and defined in that country, from which we derive our knowledge of those principles."

And afterwards, *Chancellor Kent*, 2*d vol. Com.* p. 380, adopts the language of "Du Ponceau on Jurisdiction," reading as follows: "The courts (federal) can not derive their *right to act* from the common law. They must look for that right to the Constitution and laws of the United States. On the other hand, the common law, considered merely as a *means* or *instrument* of *exercising jurisdiction*, does exist, and forms a safe and beneficial system of national jurisprudence." Should we, therefore, adopt the constructior given to the 8th of Anne, by the House of Lords, anc also in Beckwith *vs.* Hood, *K. B.*, that the acts of 1790, 1802, and 1831, give a *statutory right*, can th federal courts, like the English tribunals, take juris diction without express enactment by Congress? I England, the author or inventor has his action on th case for damages; here, the *statute* in terms grant no such right of action, except in the case of *manu scripts;* and if the right is inferred, can the Circui Court take cognizance under the act of 1819? We d not feel bound to admit the statute as the origin of a author's right to a property in his production; but i it should prove otherwise, and the federal court shoul maintain a claim to jurisdiction, in an action on th case for damages, are citizens *bound to proceed in the court exclusively* of others?

Concerning the *exclusive* exercise of *legislati* power in the Federal Government, the extent of suc authority is well defined by the Supreme Judici Bench of the Union, in Sturges *vs.* Crowningshiel 4 *Wheaton*, 193. "The mere grant of a power b Congress does not imply a prohibition on the State to exercise the same power." And, in Houston *v.* Moore, 5 *Wheaton*, 1, "The mere grant of a powe in affirmative terms to Congress, does not *per se* transfer exclusive sovereignty on such subjects." "The doctrine of the court is, that when Congress *exercise* their powers upon any given subject, the States can not enter upon the same ground and provide for the same objects." And the well-known general rule is this: where a grant is made to Congress, and in express terms prohibited to the States, or a power is bestowed upon the General Government exclusive by the letter of the Constitution, or a grant is proved, in which, *from its nature*, the exercise of the same authority in the States would be incompatible with the use of it by Congress, and Congress have exercised it in all these cases, the States have no remaining power.

Admitting that Congress *might* have invested the United States Courts with jurisdiction of cases *for damages* claimed by authors, has it been granted? Perhaps it was discretionary with the National Legislature to give such further *security*, and in their own courts; but they do not seem to have done it; and if they had, could *exclusive* cognizance thereof have been claimed by them?

The question, whether the States may or may not promote science and the useful arts, by *still further securing* to authors and inventors exclusive rights, is yet unsettled. It is clearly asserted, by Chancellor Kent, that they may to a certain extent. But the claim of a power to legislate by the States, and thus to provide encouragement, which Congress can not or will not afford, is quite different from the retention of an original right in a State to protect the existence and preserve the use of the property of its citizens. The latter, if yielded, was a sale of natural right for the purchase of political security; and the strongest and clearest evidence of the exchange is to be demanded.

It must have been parted with, if at all, on an under standing that the General Government, empowered b the grant to *secure*, would faithfully attend to it duty and guard the rights of the author, as the State had heretofore done. The former grant involves nothin but policy. The people of the States truly imagine that an uniform rule, embracing at once all the territor of the United States, would be preferable to the nu merous and varied laws of the local Legislatures They, therefore, gave Congress the power to secure for limited times, the rights of authors, etc., in suc manner, and by such penalties, as they should thin wise and proper. Let us conceive of a period whe special protection to this species of property shall b unnecessary. Since the year 1789, half a centur has passed away. The art of book-making has be come a *business*, like the manufacture of merchandise In the times of Queen Anne, learning had just escape from the walls of the monasteries, and a calling s elevated did not promise to become common amon men. A particular encouragement was given to th production of books, as had often been the case wit other commodities; and the public, after the autho was well compensated, had the work at cost, by a re version or resumption of the special protection. Time had not so far changed when the Government of th United States first considered the subject, eighty year afterwards, as to render encouragement useless. Bu we, at the close of the succeeding fifty years, are ac customed to think the human genius equal to th accomplishment of such things, unaided by law. An it may be asked, even now, whether the author desire more than the ordinary protection of his property?

In such a state of things, would it be necessar for Congress to repeal its laws, in order to plac writings and inventions *within the guardianship* of th

State law? If it *would not*, may not the author now rely at discretion upon the ordinary remedies of local courts, neglect to procure his copyright, and bring an action for damages when his work or invention is published by another without consent? We have seen that it is at least doubtful whether he can prosecute for such an injury in a civil form, except for the penalty *while holding* a copyright. In most cases, an ordinary remedy would be preferable to one-half the penalty. If it can be pursued in the local courts, it would be decidedly more advantageous. The uncertainties and technicalities of patents would be avoided in the case of inventions. A course of State legislation would be unnecessary, for the State Courts have already jurisdiction, if the Federal Courts have not. The odious exclusion of foreigners from the protection of law, as recently enforced according to the letter of the statute, in the case of the "Phantom Ship," Marryatt *vs.* Collyer, would be done away, and the necessity of an international copyright law avoided.

If, then, we have fully examined the premises, the following points are unsettled, and worthy of attention by those concerned:

1. Whether the Constitution vested in Congress an *exclusive power of legislation* over writings and inventions.

2. Whether there existed a common law right prior to the Constitution, capable of enforcement in the State Courts, without legislation.

3. Whether the laws of Congress are, at present, the sole basis of right, in property of that kind.

4. Whether the lawful jurisdiction of the Federal Courts is co-extensive with the existing rights of authors.

5. Whether it is exclusive.

Here is a broad and untrodden field of inquiry and

litigation, which the rapid advance of literature may soon render it necessary to cultivate. Some instruction on these points may be drawn from the legislation of Congress, and the action of the Federal Courts, in the matter of *patents*. The grant of the Constitution is to "authors and inventors," and, therefore, the control of the General Government is the same over each.

The first patent law, dated April 10, 1790, section four, gives to the injured patentee his actual damages, and forfeits the machine to him.

February 21, 1793. This law was repealed, and an infringement of the patent visited with damages, equal to treble the price of the invention, as sold to other persons. See section five. By section seven, if patents had been granted by the States, before the Constitution, they must be surrendered, in order to have the benefit of the act. Jurisdiction is conferred upon the Circuit Courts of the United States.

An act of the year 1800 (April 17) enlarges the law of 1793, in regard to persons. By the third section of this statute, the fifth section of the then existing law is repealed, and treble the actual damages given to the patentee or his assigns, who is injured. The Circuit Courts take jurisdiction.

On the fourth of July, 1836, another law came into existence, which in the fourteenth section, provides, that the court *may render judgment* for a sum *not exceeding three times* the actual damage, as found by the jury.

Here, as in the case of *manuscripts*, in the copyright law, Congress allows the recovery of damages in the Federal Circuit Courts, and adds a severe penalty. If these enactments are considered as bestowing damages as *separate* from the penalty, it is *competent* for Congress to give to the authors of maps, charts, etc., a right of action entirely civil and compensatory in its character. In fact, the law of manuscripts can not well be viewed in a different light.

In reference to the extent of the judicial power of the Courts of the United States, something may be gathered from the following decisions:

In the Bank of the United States *vs.* Deveaux, *et al*, 5 *Cranch*, 85, Supreme Court, 1809, it was agreed, that the *judiciary act* conferred no jurisdiction on the Circuit Courts arising from the *nature of the case*, but only from the *character of the parties;* and when citizens of *the same State* could enter that court, it must by under claims deduced from grants of lands by different States. And prior to the law of 1819 above referred to, as distributing the jurisdiction in copyrights and patents to the several courts, a case occurred in the year 1811, upon an infraction of the rights of Fulton and Livingston, relative to steamboats on the Hudson, of great hardship, where the Circuit Court for New York *refused an injunction*, on the ground that neither the *patent law* or the judiciary act specially conferred upon them *equity powers*. This was in Livingston *vs.* Van Ingen, *Hall's Am. Law Journal*, 56; and the Court say, "There being no law conferring on this court a right to take cognizance, as a Court of Equity, of cases of this nature, between *citizens* of *the same* State, our opinion is, that we can not entertain the present bill." In this spirit, the Supreme Court have ever *disclaimed* jurisdiction, unless specially given by the Constitution and laws pursuant thereto. And the subordinate Federal Courts are equally cautious in assuming judicial authority.

Recurring, then, to the case of authors, is it probable that the Supreme Court of the United States would entertain a *suit for damages*, brought by a copyright holder, for books, maps, etc.? If they would not, where is the law bestowing such a power upon the Circuit or District Courts? Again, is the instance here given cognizable by the United States Judiciary, aside from the statute, or in the absence of a copy-right? that is, have any of the Federal Courts a claim to enforce the common law right, by a common law remedy? Granting that they have not, the importance of our first query is manifest. A negative answer to that throws a duty at once upon the State Legislatures; and we are disposed to claim from the State Courts a point, which the answer to the second query will settle. If the third question is decided in the affirmative, more legislation is due from Congress, as it is also if the fourth is found in the negative. But if the negative of the fifth and more important query is true—also, the affirmative of the second, and the negative of the third; it rests alone with the authors themselves to claim of the local courts that justice which is guarantied to them, in all cases of property, by the State Constitutions.

A STATEMENT OF ELEVATIONS IN OHIO,

With reference to the Geological Formations, and also the Heights of various points in this State and elsewhere.

[American Journal of Science, Vol. xlv, No. 1, 1843.]

In giving the levels for Ohio, it should be understood that they have been taken with reference to Lake Erie, as a zero. The surface of Lake Erie has generally been considered as five hundred and sixty-four feet above tide-water at Albany; see the Report for Michigan, 1839–40. The topographer of that State, S. W. Higgins, Esq., puts it at 565,333 feet. If this last number represents the levelage of the Erie Canal, it is probably good for the surface of the Lake, as it was when the surveys were made for that work, twenty-five years since. The surface, however, fluctuates in the extreme about six feet, thus rendering all measurements based upon the Lake as a starting point, liable to an error of that amount. The Ohio Canal was explored in 1824–5, and of course its elevations are noted with regard to the stage of water at that time. The difference between 1816 and 1824 in the surface of the Lake, will render all our levels along the Ohio Canal, when referred to the ocean, inaccurate by that amount. In the latter part of the year 1815, and all of the year 1816, the Lake was *high*, about four feet above the point of greatest known depression. From 1819 to 1822 it was *low*, and in 1825 was still but about two feet *above* the lowest known point. The error in adopting the Lake surface, in 1824, as a starting point, may therefore be two feet, making its general surface above the tide-water at Albany, in 1824–5, five hundred and sixty-three feet, and at the time of the great rise in June, 1838, five hundred and sixty-seven feet. In estimating the heights given below, I have used the commonly received number of 564, to express the surface of the Lake. Both upon the Erie and the Ohio canals and other works, the slopes sometimes given to the bottom are rejected, because unknown: where there is more than one summit they counteract each other in some degree. Where there are fractional feet they are rejected. In some cases there are short intervals not measured, or the minutes of a portion of the heights are wanting, or the authorities are contradictory; these are designated by an interrogation, and will go for what they are worth. Information derived from so many sources, and transcribed many times from one note book to another by different persons, must of course be subject to errors. But it has been drawn from the best authorities, viz. the profiles and reports of the engineers in the public employ.

For location of points I have adopted Columbus, the capital of Ohio, in latitude 39° 57′ north, longitude 83° 3′ west, as the center of reference. The general course and distance from Columbus being given, the courses and distances of the different places among themselves may easily be found.

The order of stratification in Ohio is as follows, beginning at the lowest of our explored rocks, the limestone.

1. Limestone; thickness unknown, not exceeding 1000 feet; subdivided as follows by Dr. Locke: (1.) Blue limestone and blue limestone marls, over 500 feet in thickness; (2.) Marl, 25 feet; (3.) Flinty limestone, 52 feet; (4.) Marl, 106 feet; (5.) Cliff limestone, 89 feet. This limestone is the surface rock over about two-fifths of the western part of Ohio, and extending into Indiana.

2. Bituminous slate, or black shale, 250 to 350 feet.

3. Fine-grained or Waverly sandstone, 25 to 350 feet.

4. Conglomerate or pebbled sand-rock, 100 to 600 feet.

5. Coal measures, say 2000 feet.

FORMATION No. I.—*Elevation of some points at the surface of the limestone formation and at the bottom of the slate.*

Place.	Course and distance from Columbus.	Height above the ocean.	Local dip per mile.
		FEET.	
Columbus,		761	S. 81° 52′ E., 22¾ ft.
Bloomingville, Erie Co., deep cut of railroad......	N. 8¼ E., 100 m.	724?	
Dublin, Franklin Co.......	N. 30° W. 11 "	831	
Bainbridge, Ross Co........	S. 12½° W. 52 "	741	
West Union Adams Co.	S. 15¼° W. 80 "	934	S. 80½° E., 37. 4 ft.
Three miles S. E. of Dayton, bottom of cliff limestone	S. 76¾° W. 62 "	868	N. 14° E.. 6 feet.

With the exception of Dayton, these locations are at or near the outcrop of the overlapping slate, and consequently in or near the line of bearing.

No. II.—*Points on the surface of the black shale and under face of the fine-grained sandstone.*

Place.	Course and distance.	Height.	Dip.
		FEET.	
Newburg village, Cuyahoga County..................	N. 41° E., 154 miles.	764	S. S. E., very slight.
Sandusky township, Crawford Co. (east line,)......	N. 7½° E., 62 "	948?	S. E., and slight.
Big Walnut Creek, National Road................	N. 84° E., 8 "	804	nearly E.,—about 30 feet per mile.
Head of Paint Creek Canal, Ross Co................	South, 43 "	814	S. 83° E., 31. 99 feet.
Morse mill, on canal, near Portsmouth.................	S. 3¼° E., 83 "	518	

The last station is about fifteen miles east of the outcrop, which accounts for its being lower in natural level than the others. This formation occupies a narrow belt of about twenty miles in width along the Sciota valley, widening as it extends northward to the Lake. It is here about sixty miles in breadth, east and west, and extends eastward in form of a narrow strip along the southern shore, to and beyond the State line.

No. III.—*Points on the surface of the fine-grained sandstone, corresponding with the inferior face of the conglomerate.*

Place.	Course and distance.	Elevation.
		FEET.
Old Forge, Portage township, Summit County,	N. 45¾° E., 110 miles.	908
Near Chagrin Falls, Cuyahoga County	N. 39¾° E., 153 "	894
Newton Falls, Trumbull County	N. 51¾° E., 140 "	898
Narrows of Licking, Ohio Canal	N. 77° E., 44 "	764
Jackson township, Jackson County, near J. Stinson's	S. 16½° E., 55 "	785

The fine-grained sandstone region immediately succeeds the slate, and occupies a tract similar in form, though not quite as extensive. Next to it, on the east, the conglomerate is the surface rock, and from a narrow strip at the south, enlarges, after passing the line of the Reserve, to a width of fifty miles, spreading over the north-eastern counties.

No. V.—*Some points in the lowest bed of coal.*

Place.	Bearing from Columbus.	Elevation.	Dip.
		FEET.	
Brookfield, Trumbull Co., near State line	N. 55¾° E., 164 m.	990	Nearly S.,—about 20 feet.
Tallmadge, Summit Co....	N. 45¾° E., 112 "	1069	S. 33¾° E., 18½ feet.
National Road, between Jacktown and Gratiot...	N. 85° E., 38½ "	1014	S. 87° E.,—about 35 feet per mile.
Lick township, Jackson County	S. 22¼° E., 59 "	760?	Eastly and variable.

A line drawn from the Portage summit north-easterly, and parallel to the Lake shore, will be a general boundary of the coal region on the north in Ohio; and continued from this summit to the Licking summit, and thence south to the river; it will form the western limit of this great field extending to the Alleghanies.

Elevation of places in Ohio.

PLACE.	ELEVATION—FEET.	SURFACE ROCK.
Little Mountain, Lake County	1164	Conglomerate.
Mantua, Portage County, summit of Chagrin and Cuyahoga rivers	1140	"
Mahoning summit, Champion, Trumbull County	908	Fine-grained sandstone
Brookfield, Trumbull County	1154	Coal measures.
Portage, Summit Lake	958	Top of Conglomerate.
High land adjacent	1150?	
Ravenna summit, Penn. and Ohio Canal	1068	Coal measures.
Hanover, Columbiana County, Sandy and Beaver Canal summit	1123	"
Huron summit, swamp, S. E. corner of county	978	Near junction of slate and limestone.
Harrisville, Medina County, Kilbuck summit	901	Conglomerate.
Tyamochtee summit, T. 5, S., R. 15, E. Marion County	898	Limestone.
Blanchard's fork of Anglaise, 2¾ miles E. of Fort Findlay, Hancock County	1052	"
Loramies summit, Miami extension canal	942	"
Somerset, Perry County	1159	Border of coal measures.
Zanesville—river, at bridge	679	
" hill, E. of town	801	Coal measures.
Hillsborough, Highland County	1124	Limestone.
Greenville, Darke County	1044	"
Summit between Sciota and Mad rivers, near Mechanicsburg, Champagne County	1007	"
Summit of Great Miami and Sciota, Logan County	1350?	"

Height of places in Michigan, above the ocean.

	FEET.
Head waters of Belle River, Lapeer County	992
Summit between waters of Saginaw Bay and Lake Michigan	673
Pontiac summit, Clinton and Kalamazoo Canal	914
Hillsdale County, seven miles east of Jonesville	1211
Summit of Central Railroad, on the line between Jackson and Washtenaw Counties	1015
Fort Holmes, Mackinaw	797

Height of Lakes.

	FEET.
Ontario	232
Erie	565,333
St. Clair	570,005
Huron and Michigan	578,008
Superior	596,180

Height of points on the Ohio.

	FEET.
Pittsburg	705
Marietta	567
Portsmouth	479
Cincinnati	432
Summit of Wabash and Erie Canal, near Fort Wayne, Indiana	810

	FEET.
Summit of Chicago Creek and Illinois River	595
Portage, Fox, and Wiskonsan Rivers, Fort Winnebago	669

Elevations in Pennsylvania.

Conneaut Lake	1074
Alleghany summit, northern route of Schlatter's surveys	2002
Sugar Run summit, two miles north of Portage Railroad summit	2183
Chestnut Ridge, National Road	3612
Keyser's " " "	2843
West Alexandria	1797
Washington	1400

Elevations in New York.

Chatauque Lake	1291
Frankliuville, Chatauque County. (in a valley)	1588
Summit between Elm Creek and little Valley Creek, Cattaraugus County	1725
Summit between Big and Little Valley Creeks	2180
" " Cayuga Lake and Susquehanna River	981
" " Seneca Lake and Chemung River	890
" " the sources of the Alleghany and the waters of Lake Erie, at the lowest pass	1200
The ledge of *Niagara limestone* causing the cataracts of Niagara, Genesee, Oswego and Black Rivers	551
Height of land between Buffalo and Lewiston	640
" " between Buffalo and Lockport	590
Mohawk, at Little Falls	385
Hills adjacent	1097
Round top, Catskill Mountains	3804
Summit on Welland Canal	6241

This collection of altitudes is now published with a desire to bring out similar statements from other quarters. Topographical geology is of the highest value in reducing the science from a state of general calculation to the exactness of mathematical rules. But such results require great labor, and make little show on paper. If the engineers and geologists of the United States would combine and assist each other by publications similar to the above (excepting its errors), the difficulty of collecting such facts would be done away with at once. This journal appears to be the most convenient organ of such publications. When tables can be formed showing the elevation, extent, thickness, and dip of the great formations in the individual States, topographical and geological models may be constructed, which shall be miniature copies of each.

A DISSERTATION UPON THE ORIGIN OF MINERAL COAL.

[M. C. Younglove, Cleveland, 1845.]

INTRODUCTORY REMARKS.

The following argument against an opinion almost universally received by geologists was composed for publication in the *American Journal of Science*, and was forwarded to the editor for that purpose in 1842. Professor Silliman respectfully declined to insert it, contradicting as it does his own opinions.

While traversing the State of Ohio some years since, I found that our citizens took a lively interest in whatever relates to geology and the earth, and read and reflected upon what is published concerning them. Numbers of intelligent men disputed with me the doctrine of the books, that bituminous coal is of vegetable origin. A theory which met with such general disbelief, even though among men whose pretensions to science were nothing, must at least possess something unnatural. Geology itself is little more than an abstract of the *observations* of individuals; a collection of facts, by the labor of thousands. Every person has to do with the materials of the surface of the earth, and almost every observing man examines these substances with scrutiny, and naturally reflects upon the changes they have undergone. No subject of a scientific kind is more popular or better adapted to the circumstances and capacity of the mass. When a theory in such a science, therefore, meets with almost universal discredit, this is a reason for *re-examination* by men of science; although it may not of itself go far to overturn the doctrine.

It is certainly better, if it is desirable to distribute "learning among men," that the deductions of books should contradict the popular belief and common observation as little as possible. To reconcile discrepancies between what is called common sense, and science, rather than strive to widen the distance between them, by advancing what is mysterious and incomprehensible.

With these views I began a review of the doctrine of the books: that coal is of vegetable and not mineral origin; and my researches and observations have converted to me to the popular side, in opposition to the authorities. Some of my reasons will be found in this paper.

I am aware that to professed geologists my presentation of the subject may not be satisfactory. Like myself heretofore, they are now committed to the vegetable theory, and have a bias in its favor. Many of the professors in geology are in advanced life, and commenced their studies while the science was in its infancy, and when the authority of books was almost supreme. Its ablest authors have reaffirmed the doctrine in their books with new arguments and facts. Dr. Sangrado considered this a sufficient reason for maintaining his old opinions.

Perhaps some of the fathers of the science may here find enough to induce a re-examination of the old doctrine; and with them many of the younger geologists.

If this is done in a spirit of candid enquiry, untrammelled by habit and ancient authority, and my ideas should still be regarded as erroneous, the old theory will thereby receive strength, and its friends will occupy their position with new evidences of its correctness.

STRUCTURE OF THE EARTH—ITS LAW OF HARMONY.

The outer masses or crusts of the earth are composed principally of earths, alkalies, and oxides, sometimes in combination with acids or simple substances, such as sulphur, and phosphorus, and sometimes not. We have these facts by observation. They are not deductions of science, but matters of plain every day experience. The learned may know most of the properties and chemical composition of the globe on which we live, but all men have knowledge enough to observe a palpable difference between it and the animals and trees which exist upon it. They call the earth a *mineral* "kingdom" to distinguish it from the animal, or vegetable kingdoms. Geologists make use of the same terms. It is not singular, therefore, that the mass of mankind are taken by surprise, when they are told that in this grand division of the universe called *mineral*, there is *one exception;* and that an important portion of the Secondary rocks are *vegetable* rocks. Is not this apparently an arbitrary violation of that conspicuous uniformity, law, and order, which pervade the universe? Would it not be safer, more philosophical, and even more reverent towards the Creator himself, to forego the adoption of a theory so contradictory to the spirit and harmony of his works? Is it not better to be without a theory, than to adopt one which is absurd, without the clearest proof? The doctrine we are combating was one of the earliest speculations of geologists; while as yet the science was enveloped in the darkness and errors of its first dawn. Mr. Lyell says of the geological fathers, that they "indulged in many visionary theories." If the bias of their minds has been to sustain their theories, without careful examination as original questions, the authority of the later works taking the same side, is thereby much weakened. Does not the exact transmission of language and thoughts, from the early writers down through all the subsequent works to our time, prove the supposition of such a bias?

Let us suppose the question here started for the first time. That there had hitherto been no theory and consequently no prejudice—that men, content with an observation of the rocks, their order, composition, and economical value, had not speculated so much upon their origin.

On the surface they find, first, the drift or diluvium, composed of loose fragments of rocks, constituting gravel, clays, marls, and sand, confessedly of a *mineral* nature.

Descending in the natural order, next come the tertiary formations, made up of stratified beds of clay, limestone, and other rocks. Still farther down in the body of the earth, we strike the cretaceous group, with lime as a base—then the green sand, composed of earths and alkalies—the oolitic system: of alumine and lime—the red sandstone, lime and silex. Nineteen-twentieths of the coal formation itself is shale, sandstone or limestone. It rests upon the conglomerate, all silex—that again upon 1500 to 2500 feet of aluminous, calcareous, and silicious rocks, which, whatever may be their character or origin, we attribute to one method, and call them all by the name of mineral substances. At the bottom of the Cambrian system, a different method of formation is observed; no more organic remains—no farther stratification, or order of super-position; but the same ingredients are still present, lime silex, alumine and oxides of metals, even metals themselves, and they are here as elsewhere denominated minerals. In fact the younger rocks are considered as composed of the materials of the older, merely changed in position, and in form. The agent employed in this work of change is thought to be water. It is supposed to have possessed, by means of heat, or acids, or some corrosive mixtures, an extraordinary solvent power, and strong chemical force, sufficient to effect the dissolution of the ancient or primitive rocks. Yet it is denied that this menstrum could liquify bitumen, and therefore we must look for some cause for its production on the spot where it is found. That not being soluble in water, it could not have been taken up with the materials of the secondary rocks; and being specifically lighter, could not have been deposited by that fluid. It is moreover advanced as a geological truth, that bitumen did not exist until the era of the secondary rocks, because none of it is found in the primary. If it is settled by observation, that the primitive rocks do not contain bitumen, it would readily be accounted for by the igneous origin of that great portion of the earth's crust, during which the heat would consume or drive it away. The heaviest masses of bitumen are in Trinidad, Palestine, and the Burman Empire. Their relation or contiguity to primitive rocks, does not seem to have been thoroughly understood, or even examined; neither is there any evidence to connect these immense beds of bitumen with vegetable matter.

Cases are known of its presence in the transition series, some of which will be given hereafter. If it is, as contended, a kind of artificial compound, formed from vegetable matter, in the laboratory of nature, among the secondary rocks, there would not seem to be any necessity for its existence in the transition, the bitumen of these rocks not having been provided in a form, or in quality, for economical purposes. The fact of its absence from the primitive rocks (if it shall prove to be a fact) may be advantageously compared with the same state of things, in regard to gypsum salt, and other minerals of the highest value to man which these rocks do not furnish. In truth, the argument drawn from this source is not strictly in keeping with geological principles, for the most striking peculiarity of the system, next to its stratification, is the fact, that almost every mineral has its peculiar rock, and it is a common expression to denominate a formation the "lead bearing," "copper bearing," or "tin bearing" rock. It does not follow, therefore, that coal should have been found in the primitive rocks, or that its absence would be regarded as affecting the question of the manner of its creation, but only the order as to time.

From the granite upwards, where we find a stratified rock, it bears evidence of the action of the water. By this agent as now universally conceded, the materials were disposed in planes or strata. Wide seas rolled over the earth where the dry land now appears. Approaching the era of coal formation, the conglomerate rock is deposited as a basis to the series. Sometimes a coal stratum itself rests directly upon the pebble rock, but usually fifty or one hundred feet of shales, sandstone, ironstone, limestone, or fire clay intervene. All these are regarded as aqueous deposits at the bottom of the ancient sea. Ascending, through the entire thickness of the coal series, it is not unusual to pass twenty and sometimes thirty distinct beds of coal, separated by the rocks just above named. Unless, then, the coal is a product of marine plants, there must have been, during the deposition of the coal formations, as many radical changes in the condition of the surface of the earth as there are beds of coal. Most of the fossils of that formation are *land plants* of large size, and required, if not dry land for their roots, an atmosphere for their leaves and branches.

The shales, limestone, &c., could only have been deposited beneath the surface of water, and they now occupy the same area as the coal. Upon each square rod of ground, occupied by those twenty beds of coal, there must have been twenty recedings of the universal sea, twenty returns, and twenty periods of vegetable production, on the surface of the earth. And this, long before the globe was fitted for the permanent occupation of the vegetables intended for its final state. For at that epoch, the geological structure of our planet was not complete. It lacked a part of the secondary, and all of the tertiary and diluvial deposits, since then piled upon the coal. Now, although it is sometimes advanced, that large masses of rock formations were the result of gradual accretions by marine animals, like the coral polypii, and that these furnish an analogous case, the distinction is in fact too great to admit of analogy. Whether the animalculæ originate the substance of the rock, or merely extract the mineral contents thereof from the water, thus *assisting deposition*, there is nothing *unnatural* in the process. Many are marine, and calculated to flourish beneath the waters where they are found. If, instead of mollusca and crustacea, we found the fossil remains of birds and quadrupeds, and attributed the rock formation to their agency, the cases would be parallel, and equally strange.

STATEMENT OF THE QUESTION.

My proposition is simply this: *that the coal and coal bearing rocks have a common origin.* If, therefore, we call the shales, sandstones, &c., of that formation, *mineral in their origin*, and deposited from water, then the coal must be called mineral also; and an aqueous deposit. If the sandstones, shales, and iron beds, containing myriads of vegetable fossils, are misnamed, and are in fact of vegetable parentage, then the same term is applicable to the beds of coal contained within them.

The question itself would be regarded as a matter of idle speculation, if it had not assumed a prominent place in geological books now read throughout the world.

GEOLOGICAL AND LITHOLOGICAL CHARACTER OF COAL.

To begin the comparison: the coal is *stratified* like the other rocks. Its external stratification is in fact more perfect than any other rock of the coal series, except the limestone. Its superior and inferior surfaces are more regular and parallel than either the shale or sandstone. In proportion to their thickness, the coal strata extend over greater spaces, without being cut off or lost, than any other. In the roof of a mine, the rock is seen to change in character very often. A bed of shale, in contact with the coal, is displaced by a deposit of sandstone; and these thin out and thicken up rapidly, while the coal holds one uniform thickness and quality. The iron strata are more limited in extent than the coal. The main coal beds appear (when not disrupted by force) to extend throughout the basin to which they belong—in America many hundred miles across. Of internal stratification or lamination, the slates and shales are considered as offering the most perfect specimens; and as proving, most conclusively, the fact of sedimentary disposition. The lines or planes of deposit are so distinct as not to admit of a moment's hesitation. The stratum of shale, in which this conclusive record of aqueous action is found, rests upon another; which is bituminous coal, and that upon a third, perıaps as well laminated as the first. In the case which ›ften occurs, where the intermediate bed is blue lime-ıtone, instead of coal, the marks of internal stratifica-.ion are few, the fracture being as easy in a vertical ɩs in a horizontal direction. Yet without these addi-;ional proofs, there is no hesitation in ascribing the ›lue limestone to a mineral and watery origin.

The bituminous coal is a real slate. It generally ›reaks readily at the vertical joints into rectangular ›locks. In a horizontal direction, or parallel to the ıurface of the bed, it *splits into thin lamina*, from .he thickness of paper to that of a table knife.

By examining the edge of a block of coal, these ›artings or leaves are made apparent to the eye, by ;he changes of color from brown to jet black. The ıulphuret of iron, and plates of clay, are sometimes nterlaminated. The difference in color, at the edge ›f the specimen, arises from a difference in the purity ›f the lamina. When carefully burned, the brown, ›r lighter streaks, will be found to have the most ɩshes, or earthy residue. This is a general rule. If ;he coal is quite pure, the minute planes are less ap-parent, but as it embraces more earthy matter, it be-comes less valuable, and passes through all grades into bituminous shale, which burns slowly, and retains the slaty structure after the fire has passed through.

In the mines at New Castle, Ohio, the iron pyrites take a stratified form, extending many hundred feet, of a thickness not exceeding an inch. A layer of shale, about three inches thick, occupies the middle of a bed in Lawrence County, Ohio, for many miles. The top of the stratum is different in all coal beds, from the bottom, being more impure, and consequently more slaty. In fact, cut a section down through the superior shale, into the coal itself, and examine the evidences of lamination in both, there will be found no material difference in this feature, on passing from one to the other.

It is probably a waste of time to multiply proofs of the *stratification of coal*, both internal and external.

Professor Hitchcock (in his geology, p. 54) makes the following deductions from the fact of stratification—"All stratified rocks appear to have been deposited from water. Proof—the manner in which the ingredients are arranged in parallel strata and lamina, is precisely like that of sub-aqueous deposits. By no other agent that we know of is the stratified and schistose arrangement of rocks produced."

BITUMEN, AND ITS PRESENCE IN ROCKS.

Bitumen is not soluble in water; and, therefore, it is averred, and with much apparent force, that it *could not* have been in suspension by water. And more: it is specifically lighter, and *could not* have been deposited by that fluid. If, however, it can be shown, that both these things *have taken place*, the abstract difficulty must give place to the fact. And whether a theory may or may not be found to do away with the supposed obstacle, will not be a matter of much importance, after the obstacle itself has been forced out of the way.

Bitumen, whether soluble or not, exudes from the earth in thousands of instances, in a liquid state, and in connection with water. Now, for the purpose of geological union, or even of chemical action, the great object is gained when the substances are in a state of liquefaction; for they are thus enabled to obey their affinities, and move towards the center of attraction.

In Ohio, the rocks *below the coal* are bituminous. An old salt well, in Liverpool, Medina county, sent off at first many barrels of petroleum; and this boring is in the slate formation, beneath the fine grained sandstone. The well is in a valley, probably 300 feet below the conglomerate, and perhaps 100 feet above Lake Erie. The bitumen began to flow at ninety feet from the surface. Dr. Locke (Geol. Rep. Ohio, 1839, p. 271) says: "I have noticed that the rocks *above the blue limestone* and below the sandstone are bituminous, and occasionally have cavities filled with petroleum, like thick brown oil." From the blue imestone (a lower member of the great limestone fornation of Ohio) the upward section is as follows:

Cliff Limestone and Marls	272	feet.
Black Shale or Slate	251	"
Fine grained (or Waverly) Sandstone	343	"
Conglomerate——		

The Black Shale is sometimes sufficiently bituminous to burn with a bright flame. It doubtless conains bitumen and carbonaceous matter enough in the 250 feet to constitute, if collected, a heavy bed of coal. According to the Geological Report of Michigan (1838, p. 8), there are cavities in the limestone of that peninsula (called the mountain limestone) which are filled with bitumen. The geological posiion of the rocks of Ohio below the conglomerate has been the subject of much discussion. Mr. Hall, of he New York Survey, is confident that the fine grained sandstone should be classed in the Devonian System—Am. Journal for January, 1842. If this is correct classification we have 500 feet of rocks, older han the old red sandstone, that are charged with biumen That of the shale is sometimes distributed throughout the mass in a solid form. In other place: it is liquid, and flows from the earth in connectioı with inflammable gas. I have a specimen of solid bitu men from a septaria, or calcareous spheroid, imbedde(in this formation; which, when taken out, weighe(several pounds.

The fine grained sandstone itself encloses bitumen disseminated throughout its texture.

We have then established beyond dispute the ex istence of this substance, through a wide range oi *sedimentary and stratified rocks destitute of vegetabl remains.* If, therefore, the presence of vegetabl reliquiea, in the coal series, is proof that the bitume: of that formation is derived from them, their absenc in the black shale and limestone is evidence to sus tain the assertion, that in the latter rocks, it is no drawn from such a source. Even in the fine graine sandstone, vegetable fossils are most rare. Shoul the friends of the vegetable theory abandon it as t the rocks below the fine grained sandstone, suppose to be the equivalent of the Portage and Gardeau rock of New York, the question will be confined to th animal and the mineral kingdom.

For the animal doctrine, the evidence would b found as strong throughout the upper limestone as i is for the vegetable in the coal. In order to escap a natural inference, that a mineral substance had mineral origin, a resort is had to two methods o originating the same thing; one chemical, and th other animal. But even then, how is the non-fossi iferous shale to be provided? By transportation froı other rocks? How transported except by the agenc of water? And how disseminated through the bod of a rock if not in solution or mechanical suspensioı and intimately mingled with the materials composin that rock, at the time of its deposition?

So we have a substance in existence at the Silurian epoch, scattered widely through that system, entering into the structure of its rocks, stratified with it, before the age of land vegetation, and probably prior to the existence of dry land, *ascribed to a vegetable source.*

These instances are selected from the older rocks, beneath the coal; because in regard to the bituminous shale, of the coal series, it might be claimed that the bitumen, being supplied in immense quantities, might be included in the sedimentary rocks mechanically, and by the force of pressure, driven, or injected into the pores of the rock. This, indeed, would have little to do with its origin; but might obviate, in a small degree, the necessity of solution in water. But it will never be contended that it could be forced downward through 800 feet of compact rock, of a much earlier date, though such a supposition would be more rational than that of a vegetable supply. And in that case, there must have been a complete diffusion of the inflammable principle, in an aqueous medium—a thing which is held to be impossible. The coal itself is not, therefore, the only magazine from which liquid and solid bitumen may be drawn. Having shown that it exists, in a form and place where there can be no evidence found connecting it with the vegetable kingdom, it follows, that if it is true, that some coal or bitumen has had such an origin, it is not true that *all* of it originated in that manner.

CARBON, OXYGEN, ETC., EXIST IN ROCKS.

I use the terms *coal* and *bitumen* together; for it is in regard to the *bituminous principle only* that the difficulty occurs; the carbon of coal being confessedly attributable to a mineral origin. The same is true of sulphur, oxygen, nitrogen, and hydrogen, constituting, with the metals, the mass of the globe. It is beginning to be a favorite theory, that the animalculæ of silicious sedimentary rocks, and the marine animals of the old lime rock, had a share in furnishing the materials of those rocks. By this is probably meant, not that the animals created new substances, but extracted them from the waters, by which they were surrounded. Consequently, these materials had a previous existence in solution or combination, now merely modified in form by the power of animal secretion.

They are, therefore, in reality, of another origin, that is mineral; and if for the sake of analogy, the geologist resorts to the theory of animal formation, as favoring the idea of a vegetable one, it will go but a small distance towards such a result. If it is supposed that the decaying bodies of the fossil races furnished the mass of the great lime rock, for instance, it will be replied that the analysis of those bodies will not give the material which remains, the quantity of earthy result being in comparison nothing; the preservation of form impossible; and the whole doctrine of aqueous deposition is overthrown.

The silicious and calcareous *coverings* of marine animals, doubtless, went to increase the bulk of the deposit which entombed them; but these shields, in general, correspond in composition with the ingredients suspended in the surrounding medium, and are *but a deposit* from it.

FORMATION OF COAL DESCRIBED.

The words of one author do not differ materially from those of another, when describing the origin of coal, and the distribution of the vegetable masses which are supposed to give rise to it. I take those of Professor Hitchcock, p. 262. "While the earth was in a state unfit for the animals and plants now existing upon it, it was covered with a gigantic vegetation, whose relics became entombed, and were gradually converted into those beds of coal which are now in the course of disinterment."

The evidence of the existence of tropical plants at that period, with a gorgeous foliage, comparing in profusion and beauty, with the luxuriant palms of the equatorial regions, is too plain to leave a doubt on the mind. The roofs of our mines are ceiled over with *fac similes* of that ancient vegetation, preserved in the perfection and freshness of life. These, however, are distributed mainly throughout the strata *above and below the coal;* and less frequently in that stratum itself. There is, in bituminous coal, and in anthracite also, an appearance, between the layers or lamina, resembling in some degree the compressed fragments of wood or woody splinters. The substance which remains is pure charcoal, not bitumenized, and forms only a small portion of the mass. It is light and chaffy, and bears no other resemblance to wood than the fibrous structure.

The same appearance is common in the sand rocks of the coal region. Now it is not maintained that wood did not exist at the coal period; nor that it was not entombed; nor that it was not in certain circumstances *carbonized* or charred. Such processes still go on, and from *various* causes. In *some resinous woods*, a bituminous principle being present in the tree or plant would either remain or be given out. But we look upon these facts as mere incidents which occurred then and occur now and always will occur, whilo timber grows and is subject to chemical action by acids, heat or pressure; and deny that these are great causes put in operation to build up this geological edifice, the earth.

There does not appear to be any reason why the coal plants should congregate in greater quantities in the vertical space of a few feet occupied by a coal bed than in the same space in the shale above. In fact the shale generally seems to contain as many of them as could be piled together in the same space. And by observation these stems are generally confined to the shale and other rocks being almost entirely wanting in the coal. The argument then, by the evidence, is that the shales are of vegetable origin, and that coal is derived from some other source.

SUPPLY OF VEGETABLE MATTER

And passing by this objection where shall an adequate supply of woody fibre be sought for? The imagination having no bounds to its creation has constructed a forest of gigantic growth, where towering trunks crowd against each other for want of space. Let us dismiss fancy and resort to facts. We can compare the ancient productions with that of our recent western forests.

How much more timber *could grow* at the same time upon the same surface than we now observe? To the utmost it could not exceed double the crop of the bottom lands of the west. What kind of timber was it? No fossil trunk has been found exceeding sixty feet in height. The diameters of the sigillaria, though flattened, and consequently extended in width, seldom exceed two feet. Were these trees more solid, did they contain more bituminous or carbonaceous matter than the oaks, hickorys, and elms of our for ests? In general, plants of the quickest growth con tain more of the watery and less of the fibrous material. The ancient growth, being tropical, must have been rapid.

But aside from any mere deduction, the impressions left to us, showing the space occupied by a trunk of a size equal to a large tree of our times, are conclusive to every observer that they were either hollow, or soft, spongy, and succulent plants. Stems having a diameter of one to four inches become a mere plate, like glass. Lepidodendra a foot through are compressed into the space of an inch and less. Sigillaria which when found erect are round, and have the appearance of a sturdy tree, when bent down or found prostrate beneath the weight of shale, present a mere scale of matter between the opposite bark. When the limbs or bodies of coal plants are found in a vertical posture, their form is often round and apparently perfect. Is this body transformed to coal? Very seldom indeed. Its bark is commonly represented by a coating of coal; but the body, by the shale or sandstone which surrounds it. Sometimes, especially with stigmaria and lepidodendra, where branches pass upward from one stratum to another, the stem is engorged with the material of the lower stratum for some distance above its surface. This indicates a want of solidity of the parts or hollowness which allowed the immediate injection of earthy matter. Many other facts might be added to prove that the ancient tropical ferns of the coal period were even less capable of furnishing carbonaceous matter than our modern plants or trees. The amount which the latter could have supplied will be shown hereafter. But was it not derived from the bark? or from the bodies of thin and succulent plants not containing wood? If from the bark alone, how was it separated from the bodies? Where are the naked trunks that produced it? And if it is matter of astonishment how the soil of those times could produce sufficient vegetation for a supply on the supposition that *every part* was worked into coal, in what regions shall we seek for that incredible fertility which could furnish a sufficient quantity, when only an hundredth part of the vegetable is converted? When we find a stratum of iron ore, or of limestone, shale, or any other rock of the secondary class, and are told that its materials were brought from some distance, there is nothing very strange in the proposition, for we know that vast quantities of these materials *do exist* in various portions of the earth. The question of an adequate supply is in these cases easily settled. But if the limestone or the iron are ascribed to a foreign source, not mineral, to substances not found in the earth, nor naturally belonging to it, it is incumbent on those who advance such an idea, to account for the quantity of this foreign matter. Therefore, an enquiry into the *sufficiency of the supply* is a material one.

AMOUNT OF VEGETATION.

The backwoodsman knows that the heaviest growth of the western forest, with all its leaves and brush, would not bridge over the ground on which it stands, with a layer of compact timber one half foot in thickness. This reduced to charcoal will diminish in weight about eighty per cent. In this process the carbonaceous portions are retained, showing that this timber would, if converted into coal on the spot, make but a very thin plate, not exceeding and inch in thickness. The weight of coal is from *two* to *five* tenths greater than water; that of timber *one-third* to *one-half* less. After the watery and volatile portions are driven off, the weight of the carbon and the salts is about twenty per cent. of the timber from which it came. A bed of coal four feet thick would require, on the supposition that the ancient vegetation was double the strength of the modern, and the vegetables grew upon the same ground, twenty-four successive crops to complete the stratum; and this without furnishing anything but the carbon, and not the bitumen. From very few modern trees can that substance be obtained in quantity. Twenty successive beds, of the thickness of four feet, would therefore call for 480 times the timber which can grow upon the soil at any one time. The evidence in proof of transportation from abroad, as well as reason and analogy, is adverse to such a conclusion. If the growth was contemporaneous with the deposit, it must have taken place upon higher land. How could it be reached by the waters? If reached, how could vegetation continue? Brought from a great distance, how could the most delicate fibres of a tender leaf have been preserved as we find them? Not an occasional instance, but almost universally. And the minute markings on the exterior of the coal stems, is it possible that these should have survived in all the perfection of original growth? If the currents were rapid, certainly not; if they were moderate, the whole must have perished by exposure during a journey of many hundred miles, or at least the leaves, bark, and softer portions.

In the great coal field lying at the western base of the Alleghanies, the northern and southern axis exceeds 500, and the other 150 miles in length. And suppose, instead of twenty strata there was but one, each square mile demands the product of twenty-four, or for one bed of this basin more than 1,000,000 of miles, a space greater than the valley of the Mississippi, from which this growth is to be gathered in.

In a broad sea, what shall create a current from all sides to the centre? What mysterious agent would guide the floating mass to every part of the field in equal quantities, forming beds of equal thickness, carry it to the bottom in regular and compact order, free of the sediment of those waters, and in the next moment clothe it with a heavy mass of the same sediment?

TRANSPORTATION OF TIMBER.

The Red River has its rafts, extending many miles in length, which float upon the water. The timber is detained by the narrowness of the stream, its upper portions decay, and its lower sink piece-meal to the bottom.

The Missouri and Mississippi, broader and more powerful streams, bear their floodwood in solitary trees to the ocean. Sometimes it descends to the bottom, or is thrown upon the shore and covered with mud. Here are the prominent cases known to us,—the very cases cited by geologists to sustain their position, furnishing nothing analogous to the process required in the formation of coal. Here we have the timber and the currents of water, the sediment to shut out the atmosphere and set the chemical agents at work. The timber sinks in a confused mass, the interstices are filled up with mud, and, instead of carbonization, there is either rottenness and destruction, or, thus embalmed, they are preserved in kind. But waiving all other objections to the supply, transportation and deposit of timber, the fact of internal lamination in coal is wholly incompatible with this doctrine.

ANALOGY OF PEAT.

Peat bogs have been called in to aid in the manu-ıcture of bituminous coal. No *stratified* beds have ɜen found supposed to proceed from this source, for ɜat itself does not exist in strata; but carbonaceous atter, allied to coal, has been observed; and trunks ' resinous trees surrounded by it have become car-ɔnized. From this the inference is taken that coal *ight* have been formed under these circumstances. ow although it is something of an argument that a ıing *might* be and not outrage any law of the uni-ɪrse, it is a much stronger one that it *could not* be ıless such a violation has taken place. Peat bogs) not and can not flourish under water. They are e product of vegetation in the northern latitudes; e result of cold and not of heat. The coal plants e all tropical. Peat is mere alluvion and recent, ɪver having been found in quantity embraced in cks. It is an irregular, limited, and accidental de-ɪsit, having in form, composition, and geological lation, about the resemblance to the coal strata, at concretionary lime or calcareous tufa has to the eat limestone formation.

TRANSFORMATION OF WOOD INTO COAL.

Surturbrand, Lignite, Bovey coal, and the beds ar Munden, are other instances brought forward to ove that it was from real trees that the coal beds the earth were derived. Here it is said that the rms of timber exist, and are well preserved. It is, believe, from these beds that Mr. Parkison selected s specimens exhibiting the fibrous structure,—an periment which appears to have satisfied mankind of the fact that all coal has the same woody rep sentations. These, however, belong to the rece formations above the coal-bearing rocks, the tertia diluvium and alluvium. They are only found in lim ed heaps, scattered here and there like the peat bo just noticed, and are not real coals.

At length Mr. Hutton has made a microsco examination of the Newcastle coal, and testifies the fact of an arrangement resembling the vegeta structure, with distorted pores or tubes, filled with more volatile bitumen than that of the body of t coal. I have before remarked that it is quite sing lar that *no more* vegetable fossils are found in t coal strata and in the limestone strata of the c formations. If they did not grow upon the spot, l were brought from abroad, it would be incredible tl these strata should not have enclosed many of the If the productions of the immediate vicinity, th absence at the period of the formation of each pa cular stratum may be accounted for by supposin cessation of growth, or an unfavorable influence up vegetation by the material there held in soluti The impression of plants and trunks of trees is wl we should expect to find in all the strata of that riod, the coal as well as the sandstone.

Upon the theory of mineral deposition th *should be* at least an occasional fossil, and such find to be the case, but not in so great profusion as warrant the inference that the mass was composed them. The limestones of the coal region often hibit the same scarcity. Why, then, should the p sence of a fossil tree, with its bark, fibres, rays, a concentric layers, *of itself alone* furnish any m proof that such was its origin, than it does of t same fact in the sandstone? Was that rock e thought to be a vegetable product on that accou the shale in which the growth of that period is de-sited *for that reason* attributed to such an origin? l the evidence arising from the mere presence of ınts applies to the shales with multiplied force. It s also been observed by geologists that the shale mediately above a coal bed is much more bounti-ly supplied with trunks and plants than any where e, lying in immediate contact with each other, ›ssed and intermingled in a thousand directions. e weight of authority is in favor of a vegetation on or near the spot where they now lie; and the ısequence drawn from their existence by the au-ır is, that this surplus of the ancient herbage was mineralized. It appears however to be as fully ıeralized here as elsewhere, and to have grown as fusely as it is possible for it to vegetate with suc-s, to have been buried under the conditions required form coal, and yet this compound is not produced. find only the bark and leaves so converted,—the nks represented by shale, and the interstices, as should naturally expect, occupied by the same k. Now may it not be supposed, with more rea-, that the coal being already deposited presented ho plants of the succeeding stratum a new stimu- prompting them to a rapid growth. That each tum has its peculiar botany is well known. The clay beneath the coal seldom has any thing but stigmaria, and always a plenty of that. This ıt is a heavy *marine* vegetable, and not belonging he land. It is not even supposed to have been d, but grew floating from place to place. What ing is there for terrestrial plants while the stig-ian stratum or fire clay is in formation. The ninous and silicious shales, overlaying the coal, their flora much more abundant and various than solid sand-rock, because a more moderate and t deposition favored such a result.

COMPOSITION OF WOOD AND COAL.

But admitting that all coal strata will, on ex mination, exhibit the marks of woody structure,· branches and leaves,—if from this fact, (should prove to be one) we step to the conclusion that the trees and vascular plants contained *either the quant or the kind* of the elements which remain, we sh be in error. Unless the body of each individual t or plant contained within itself the ingredients coal, the deficiency must come from some other t or source. Derived. in this manner, there must b decomposition in order to receive the foreign ing dient. And if the woody fibre contains substan *not in the coal,* a like process must have been pas through in order to allow them to escape. Suc transformation would imply a chemical dissolution recomposition, and this would certainly destroy texture of the wood, though the form might be served. If it shall be affirmed that the parts not stroyed are filled by the results of those that are, are forced to adopt the conclusion that, under effect of the same agents acting together upon same substance, only a part yielded to the forces plied, and a part did not, the consequence of would be that the portions not overcome would re not only their original form, but also their *nat* unchanged.

An analysis of five English coals gives,

Carbon	72,23
Hydrogen	9,44
Nitrogen	9,79
Oxygen	10,26

Three others give,—

Carbon	77,53
Hydrogen	3,76
Oxygen	17,24
No nitrogen.	

Three of the Ohio coals give the following re-ults,—

Carbon	47,745
Bitumen	40,004

Four specimens from George's Creek, Maryland, ave,—

Carbon	81,37
Bitumen	18.37

Mr. Turner, in his work on chemistry, considers aphtha, petroleum, mineral tar, and other liquid itumens, as composed essentially of the same mate-ials. According to Dr. Ure, naptha contains,-

Carbon	83,04
Hydrogen	12,81
Oxygen	4,65
Nitrogen . . . . . -	None.

On the supposition, in the absence of analyses, hat the bitumen of the Ohio and Maryland coals is imilar to Naptha, we shall have for Ohio, carbon—0,945; George's Creek, carbon—96,66. No vege-able can furnish within itself this amount of carbon. ven pure fibre gives for carbon but 51,43. Mr. iebig (Organic Chemistry) and Mr. Buckland sup-ose a part of this given off in connection with the

11

oxygen. The fibre is only a part of the woody structure. So we have only fifteen to twenty per cent. of the mass of the tree in the form of carbon.

Fibre contains, in addition to the 51,43 of carbon, hydrogen 5,82, *oxygen* 42,73. The volatile parts of wood form acids and gums.

The gummy and resinous portions of most vegetables form only a minute proportion of this eighty per cent. of volatile matter.

Take the most favorable known case, the resinous pines, or other conifera, which produce tar by distillation, and compare the proceeds with the bulk from which it is drawn. The pine, however, is not found in connection with coal, nor is there sufficient evidence to show that other resinous trees were present.

SUPPLY OF CARBON.

From whence, then, is the carbon of bituminous and anthracite coals to be obtained? From the animal world or from the mineral? Should it be found necessary to resort to the latter kingdom for any part of the supply, all necessity for the assistance of the vegetable is done away; for the theory takes its existence and force from a supposed want of bitumen in the original structure of the earth.

EARTHY RESIDUE OF WOOD AND COAL.

Again, where are the earthy constituents of wood which should remain in the coal? In the report of the re-survey of Massachusetts it is said,—"When we look at the analysis of vegetables we find these inorganic principles constant constituents, namely, silica lime, magnesia, oxide of iron, potash, soda. Hence these will be found constituents of all soils."

In the supposed change which vegetables undergo in the process of carbonization, is the disorganization less complete than in the decay of timber and leaves which enter into the composition of soils? Coal contains neither lime, potash, magnesia, or soda. Five western soils gave by analysis,

	Sulphate of lime . .	2,32
	Phosphate " . . .	0,66
Four of these,	Carbonate " . . .	2,22 go to

the resinous trees. The ashes of the *pinusabres* contain *three* per cent. of *phosphate of lime;* and even the pollen of the same tree has the same per centage. The average phosphate of the *birch, basswood* (or linn), and *oak* is 4,83 per cent. All vegetation seems to imbibe the earths and salts, which are given back to the soils by their decay. Such a result is inevitable. The material not being destructible, even by passing through the fire, which reduces the fibre to ashes, nor volatile like a gas, necessarily remains with the residue, and if it ever entered the plant would still have been found in its remains.

In morphia, sweet almond oil, and a few other *artificial vegetable extracts*, nitrogen has been found. In the two compounds just named 5,53 per cent. (according to Dumas), but with *none* according to Dr. Thompson. Indigo and Cinchonia have nine to twelve per cent.

SUPPLY OF NITROGEN, ETC.

From whence is the nitrogen of the five English coals derived? If the guns and resins are analogous to the bitumen, which it is claimed is produced from them, they can not furnish it. The woody fibre can not, and the mineral world is present with bountiful supplies. Here it exists, and forms its combinations, why not enter into coal? So with the carbon and hydrogen. All the materials have a being in a mineral state,—are placed in contact under circumstances favorable to obey their chemical affinities; whence therefore arises the necessity for crowding aside a simple, consistent, and natural explanation to introduce another, which is marvellous and inconsistent. Probably no theory will *completely explain every thing* which is seen in connection with the coal beds. There are some phenomena in every rock which perplex the geologist. The true philosophy is to adopt the fullest and most rational theory, or none at all. It is not the duty of a geologist to speculate upon the creation, and account for every strange appearance he may discover. He is not bound to do this. It may be considered an accomplishment and a subject worthy of study, but he has high practical functions to fulfil, dealing rather with the mineral properties, the economical value of the rocks, as they exist, than with the manner of their existence. My object is to offer reasons why the coal strata should not be separated from the coal-bearing rocks, in the method of their creation, not to account for all the observed phenomena connected with them. I have referred the coal to a mineral origin, because geologists refer the adjacent rocks to that system. By it most, if not all, of the attending facts may be accounted for, perhaps as many as of any other rock; but this is a separate discussion, for which there is not room here; and by it we are not driven to imagination upon imagination in order to make out a case. We escape the inconsistency of supposing that the vegetation of the surface of the globe became necessary to the upbuilding of its mass. The incredible fancy that the unfinished earth destined for vegetation should, in this stage of its creation, produce a growth inconceivably heavier than it now does in its perfect state.

That this precipitate verdure should, while occupying the valleys and hill sides of that day, be gathered in from the wide-spread regions around, and consigned to the bottom of a sea in a separate mass. The withdrawal of that sea to allow further vegetation, its return to harvest and bring it home repeated many times.

These arguments may not be regarded as conclusive, but appear to me worthy of consideration. Geology is not in its nature a science capable of exact demonstration in many important points. It may be presumptive to question a doctrine as old as the science itself; but errors have been maintained for a longer period than this in every department of knowledge. The doctrine was adopted when but a very small proportion of the coal regions of the earth were known, by men who based their opinions upon the brown coal of the tertiary system; they may have in this, as in other cases, adopted a "visionary theory." If it is so, for the credit of the science among mankind at large, it can not be exposed too soon.

My convictions on the subject are of some years' standing; they are made public with full confidence that the facts falling within my limited observation can be reconciled in no other way. At the same time I can not say that others more learned, with more extensive resources of information (and consequently more light,) and especially with the opportunity of personal examination in different continents, may not be as much at a loss to reconcile what they have seen and known by the theory here advanced, as I have been by the doctrine of the books.

LORD DUNMORE'S EXPEDITION TO THE SCIOTA TOWNS, 1774.

[Sandford & Co., Cleveland, 1842.

Gentlemen of the Historical Society,—

In the selection of a subject for this discourse, it seemed to me necessary that it should be of an historical rather than a philosophical kind. It is evident, from the circumstances which attended the formation of this society, that it was originally intended for historical purposes only. The gentlmen who brought it into existence were not men of philosophical acquirements, though many of them were possessed of profound general learning: they were, in the old phrase, the "pioneers of the country." The striking circumstances of that undertaking were still held in vivid recollection by them, and they felt that this thorough revolution, effected through their means, was worthy of a history in detail. They found here, at their coming, nothing but a wild though luxuriant waste, occupied by a barbarian race, who drew their supplies from the spontaneous abundance of nature, always consuming, and rarely if ever producing. The Indian occupant, although he had held undisputed possession for a period of perhaps one thousand years, had formed no permanent connection with the soil. He had not fulfilled the apparent design of heaven, that the earth should become every where the scene of cultivation and permanent habitation. He roamed through the land, like the wild animals of whom he was in chase, a passing wanderer rather than a fixed resident, leaving behind him no marks of his presence save the rubbish of his wigwam and the ashes at its door. The falling leaves of autumn concealed the one, and the decay of the seasons dissipated the other. Even the places of his burial, which men love to make known by permanent objects, were without other monuments than a small white flag, rudely fastened to a pole, and a hieroglyphic record of the enemies he had slain in battle carved upon a tree. This luxuriant soil, destined for the sustenance of many millions, expended its powers in the production of an annual crop of weeds and flowers, to go annually into decay upon the spot where they grew. The riches of the mineral world lay dormant and unknown. The mechanical agency of our streams was not brought into requisition, and the great lakes, instead of being the medium of communication between opposite and distant countries, were impassable to the Indian in his light canoe. Not one of the abundant resources of nature was fully developed.

We now observe every where the exact reverse of all these things, and those men who may be considered the founders of the Historical Society were eye-witnesses of and participators in the change. The incidents attending it were necessarily interesting, and it became to them, now in the decline of years, a matter of high consequence that there should be a common repository, where each might place on record the reminiscences of his early life. And to all others, and to posterity, it is of importance that the heroic enterprise of those men, whose bravery and endurance has perhaps no parallel, even in the relations of fiction, should be authenticated and made public.

It would then be a deviation in me from the

wishes of the projectors of this institution, and doubtless of its present members, to introduce any other than an historical subject. I have therefore collected some facts relative to the expedition of Lord Dunmore into this region in the fall of 1774. With the opportunity allowed me, it will not be expected that much original information can be presented to you in relation to this campaign.

With this, as with most of the early war parties beyond the Ohio, no journalist or regular diarist appears to have been present. It is indeed probable that there may be, among the descendants of the officers and soldiers connected with those military expeditions, the remains of memoranda of dates, deaths, and other incidents, still in existence. We might reasonably expect to find, either in Virginia or England, a sketch of the operations of Lord Dunmore, prepared by himself, his secretary, adjutant, or some of the surgeons. An address to the Historical Society of Virginia, and some individual or British Society in London, would doubtless bring to light something of the kind.*

There were also several incursions from Virginia, Pennsylvania, and Kentucky, into the country now called Ohio, between the time of Dunmore's expedition and the campaigns of St. Clair and Harmar. These were often mere marauding parties of individual and not official enterprise; but at this day it is impracticable to obtain many details respecting them.

* Since the delivery of this discourse, a letter from John Connally to Colonel George Washington, dated Winchester, Va., Feb. 9, 1775, has fallen under my observation, and contains the following paragraph: —

"I have transmitted a copy of the Treaty to his Excellency, and should have sent you one also, only as I have desired the journal of the expedition to be printed including the whole, I deemed it unnecessary."

They were composed of dauntless men, who thought, fought, and acted, with great energy, but seldom or never wrote.

In 1778 Major McIntosh collected a party in Pennsylvania, and penetrated the Indian country to the Muskingum river, near Bolivar, in Tuscarawas county, and built Fort Laurens. His route was along the Tuscarawas trail, a track or highway of the aborigines, portions of which are still visible on the waters of the Little Beaver and Sandy rivers.

In 1782 General George Rogers Clark issued from Kentucky at the mouth of the Licking (opposite Cincinnati), with a small command, and proceeded to destroy the Pickawee, Chillicothe, and Wills towns, situated on the upper waters of the Great Miami. There is an intimation that this was not his first expedition against the Pickawee towns. The Wapatomaka villages on the Muskingum, near where the town of Dresden now is, were attacked and destroyed in July, 1774, by a party of four hundred men under Major McDonald.

In the year 1782 Colonel Williamson, who had accompanied Lord Dunmore seven years previous, embodied a force in Pennsylvania, and came to the Moravian towns, on the Upper Muskingum, where was committed the noted massacre of the Christian Indians. Soon after Colonel Crawford made his unfortunate advance upon the tribes and villages on the waters of the Sandusky, by way of Fort Laurens. Other expeditions are reported to have taken place from time to time, but there is very little known respecting them. I have made mention of them here for the purpose of calling the attention of gentlemen travelling in Kentucky and Pennsylvania to the subject, hoping that something of importance may thus be collected from the survivors or their families.

The meagre details which I am able to find in ou printed histories concerning the operations of Dun more in the Indian country, show us how little i known of the transactions of that eventful period.

To understand fully the immediate causes of Dun more's war, we should refer a moment to the previou relations of the Indian tribes to the French and Eng lish colonists in America.

The French established themselves at Montreal i the year 1607–8, about the time of the settlement o Jamestown, Va., by the English. Prior to the yea 1642 the "Six Nations," as the confederacy occupy ing Lake Ontario and the east end of Lake Erie wa called, had given only occasional trouble to the Britis colonists; in fact they had held amicable meeting and perfected treaties of friendship. The colony o Wm. Penn, after the influence of the French ha ceased, reposed much confidence in the northern In dians, regarding them as stedfast friends. It wa otherwise with the western and Scioto tribes, em bracing the Shawanese, Delawares, Mingoes, Tawas and Wyandots, occupying at that period the water of the Scioto and that portion of Lake Erie west o the Cuyahoga river. The French had taken possessio of their country and their friendship at an early day The Jesuit Missionaries were at Onondoga in 1656 and were found by La Salle at many places on th lakes and on the Illinois river in 1680. There ar other evidences of their early occupation. I sa during the last summer the stump of an oak tree upon the farm of Hezekiah Chidester, in Canfield Trumbull county, in this State, which had been cu by an axe when the tree was young, doubtless th work of the French adventurers. Relying upon th circles of growth as each corresponding with a year i time, this cutting was made 180 years previous, or i ıe year 1660. Axe marks were found in a white-ood, or poplar log, in Newburgh, Cuyahoga county, ɔme years since, of about the same age. These 'andering Jesuits were soon followed by traders, and ıe latter by a few ardent discoverers, each pushing is way in advance of the other, till at last, having aving passed through the entire length of the north-rn lakes, they struck upon the Mississippi, and this ɔd them to the Gulf of Mexico and the ocean.

The relations of the French government to the English being those of perpetual jealousy and con-ict, and the former power having now encircled the ossessions of the latter in America, with a chain of osts in full communication, they mutually transferred portion of their warlike operations to their respec-ive colonies here. The actual occupation of the English, in 1754, was limited on the west by the Alleghanies. The activity and enterprise of the French had so far outstripped that of the English, hat while the attention of the colonists was directed o the reduction of the French power about the Bay f Fundy and Lake Champlain, they were unexpect-dly surrounded by a cordon of fortresses, sustained y a formidable savage force. By the treaty of Aix-a-Chapelle, in 1748, the differences between the res-ective colonies of Nova Scotia and New England vere arranged; but the question relative to western ossessions were left for reference to an arbitration n the field. The western Indians, having their na-ive hostility to the English well inflamed by the fforts of French traders, officers, and priests, entered ıpon the French war with more than usual vigor. This contest continued eight years, and resulted in he complete abandonment of their Indian allies on he part of the French, and the confirmation to Great Britain, by the treaty of Paris, 1763, of all the French possessions in North America. In this condition, the western confederations, left to cope single-handed with the whites, now crowding upon their territory, seem to have hesitated for a time as to the policy it became them to pursue. They were numerous in the field, delighting in war, burning with revenge, and anxious to preserve a country to which they were attached from further encroachments. On the other hand were the sturdy colonists, indulging a thorough hatred of their red enemy, thirsting for the possession of his soil, full of courage, and capable of endurance as soldiers. War, under these circumstances, could not long be deferred; and accordingly, in 1774, eleven years after the treaty of Paris, a general conflict began, which continued—with unexampled fierceness and attended by peculiar horrors—until the victory of Wayne at the Maumee Rapids, in August, 1794. It is of the leading expedition of that war that I propose to give some account.

The immediate rupture was occasioned by murders on the part of the whites, in the following manner:

After the French claims to the lands upon the Ohio were disposed of, the colonies of Virginia and Wm. Penn set up common title to the country about the confluence of the Alleghany and Monongahela rivers. The followers of Wm. Penn, adopting the French rather than the English policy towards the Indians, had thoroughly conciliated such of them as were within their settlements, and lived with them as friends and brothers. Penn's successors were in possession. In the spring of 1774, Lord Dunmore, as the Provincial Governor of Virginia, issued a proclamation warning all the inhabitants west of Laurel Hill of their allegiance to that colony.* And an

* American Archives, 4th Series, Vol. I, p. 790.

gent of Governor Dunmore, by the name of Conolly, ontrived to get possession of Fort Pitt in the name f Virginia. The Governor of Pennsylvania put orth a counter proclamation, and the Westmoreland ilitia were mustered into service, partly on account f Indian disturbances, and also to sustain the authoity of Pennsylvania. About this time a party of Delawares and Shawanese, on account of their friendhip for the Pennsylvanians, escorted a party of their raders up the Ohio in safety to Fort Pitt. Conolly, he Virginia agent, fell upon this band by surprise, t the mouth of the Big Beaver, and wounded Silvereels, their chief. On the 23d of April, a party, eaded by Captain Michael Cresap, killed two Inians a short distance above Wheeling.* He also ttacked a camp below Wheeling the same day, and illed several. The next day, one Daniel Greathouse, is two brothers, and twenty-one men, massacred *leven* Indians at Baker's tavern, which is forty miles bove Wheeling, and nearly opposite the mouth of Yellow Creek.† This is the affair to which Logan, he Mingo chief, refers in his famous speech, supposng it to have been conducted by Cresap.

Logan was in camp opposite Baker's, and was robably among those who attempted to cross over o the tavern after the first slaughter.‡ He is known o have lost some female relatives at Baker's, and ther connections at the camp which Michael Cresap ttacked. The excuse offered by the whites for the ommission of these murders was, that the Indians

* Letter of Devereux Smith, June 10, 1774. American Arhives, Vol. I, pp. 344—467.

† Declaration of John Sappington. Jefferson's Notes, Appendix, p. 46. Also of Charles Polke, p. 26.

‡ Jefferson's Notes, Appendix, p. 44. An affidavit of James Chambers, p. 24.

had committed some late robberies, and premeditated a general attack upon the frontier settlements.*

* In the fall of 1850 the following communication was published in the *Pittsburg Gazette*, relating to the Yellow Creek murders:—

THE YELLOW CREEK MURDER AND LOGAN'S SPEECH.—This truly eloquent speech, in which the guilt of the murders near Yellow Creek is charged upon Colonel Cresap, has given to that occurrence a prominence in our history beyond that of any other of similar character. The writer of this article became early satisfied that great injustice was done to a brave man and a patriot in that admirable production, whether it was really the work of Logan or some other person, and he was led, by a desire to exculpate one who, in that case at least, was innocent, to collect what evidence he could for that purpose.

In course of his enquiries he ascertained, some two years ago, that there was living in Ohio, a few miles below Yellow Creek, a certain Michael Myers, the very man who shed the first blood which led to the killing of the Indians near Yellow Creek.

He then determined to embrace some early opportunity to obtain Myers' statement from himself, although his informant, Mr. Sloan, a respectable and intelligent gentleman, and neighbor of Myers, had often heard his story, and repeated it to the writer.

On the 21st of February last the writer called on Mr. Myers, in company with his neighbor, Mr. Sloan.

He found him a stout, vigorous old man; his memory seemed good, except in the recollection of names, many of which he had forgotten; he did not remember the names of Cresap, or Clarke, or Sappington, or Conolly, or Lord Dunmore, although he had descended the river as far as Grave Creek when that nobleman led the expedition to the Scioto in 1774.

The writer, after stating the object of his visit, asked Myers how old he was; he replied 102 years. His wife, who is an active woman, interposed, and said his age was ninety-seven; and Mr. Sloan remarked that, from data furnished him many years ago by Mr. Myers, he had supposed that he must be about the age his wife made him. This conclusion is rather corroborated by a subsequent remark of Myers, that he was a man grown when the affair at Yellow Creek happened.

His account of that affair was as follows:—

In the month of May, 1774, he went across the Ohio, near Yellow Creek, in company with two other men, to look at the country. They went up the creek two or three miles to a spring, at a place now known as the Hollow Rock, where they concluded to encamp that night. Having spancelled their horse, they turned

3oth parties being now fully exasperated, the work f retaliation began on all sides, and continued, with

im loose around the point of the hill, where there was good graz-ıg, and began kindling a fire. Soon after they heard their horse's ell tinkling, as though he was moving rapidly. Myers then sus-ected that a wolf had scared the horse, and, taking up his rifle, e ran round the point of the hill until he saw the horse standing ill, and an Indian stooping down beside him, trying to loosen the ancels. Myers immediately raised his gun and shot the Indian, ıd as soon as he had loaded again he ran up the side of the hill ıtil he discovered a large number of Indians encamped, and one ith a gun running towards him, but looking towards the horse; immediately fired at the second Indian, and, without knowing hether he had killed him or not, he (Myers) wheeled about and n towards the spring and camp, when he found that the other an had become alarmed and left before him.

Next morning several Indians came over to Baker's station to quire who had killed the two Indians the evening before, but reat House, who appears to have been the master-spirit, ordered e men not to tell, and the Indians returned to their encamp-ent.

That afternoon, or the next morning, a large canoe full of In-ıns was discovered crossing the river; the white men immedi-ely seized their rifles, and ran down to a point, where the canoe uld be likely to land, and, lying concealed until it came quite se, fired and killed every person in the canoe but one.

Here the writer inquired whether there were any women led.

Answer—"I can not tell, they were altogether in the canoe."

Such is Myers' narrative, and I have thought it worth pre-vation; of its truth every one can judge for himself. Mr. an has known Myers for about twenty years, and has heard his tement again and again, without variation, and his version of ers' narrative agrees precisely with that of Myers himself to writer.

Myers is well known as a veteran Indian fighter; his story was d without the least shadow of braggadocio, and certainly with-t any appearance of an effort to exonerate himself from a charge criminality.

He spoke of killing the Indians with quite as much indiffer-e as an experienced hunter would of killing a bear.

This narrative, if it be relied upon, certainly palliates, in some gree, the atrocity of the outrage at Yellow Creek. It refutes charge of inviting the Indians over, making them drunk, and n murdering them. It negatives the charge of treachery and remarkable cruelty, during the summer and fall. In August Lord Dunmore collected a force of 3,0(men, destined for the reduction of their towns on tl Scioto, situated within the present limits of Pickawa county. One half of the corps was raised in Bot tourt, Fincastle, and the adjoining counties, by C(Andrew Lewis, and of these 1,100 were in rende vous at the levels of Green Brier on the 5th of Se tember. It advanced in two divisions: the left win commanded by Lewis, struck the great Kenhawa, a) followed that stream to the Ohio; † the right win attended by Dunmore in person, passed the mountai at the Potomac Gap, and came to the Ohio somewhe above Wheeling. About the 6th of October, a ta was had with the chiefs of the Six Nations and t Delawares, some of whom had been to the Shawan(

infamous violation of the rights of hospitality; but still leave case of unprovoked murder. C.

* Message of John Penn, Governor of Pennsylvania, to Assembly, July 18, 1774, American Archives, 4th Series, Vol p. 602:—"I am to inform you that in the latter end of A last, about eleven Delaware and Shawanese Indians were barl ously murdered on the river Ohio, about ninety miles below Pi burg, by two parties of white men, said to be Virginians. soon as the unfortunate affair was known on the frontiers of province, messengers were dispatched to assure the Indians t these outrages had been committed by wicked people, without knowledge of any of the English governments, and reques that they might not be the means of disturbing the friendship isting between us. This step had so far a good effect as to q them for the present, and to prevent them from coming to a r lution to enter into a general war with us. It did not, howe restrain the particular friends and relations of the deceased, v it seems, contrary to the advice of their chiefs, in a short t afterwards took their revenge by murdering a number of Vir ians, settled to the westward of the Monongahela."

See also letter of Wm. Preston, of Fincastle, page 707, d Aug. 13, 1774, detailing the Indian murders in his vicinity.

† Letter of an officer from Fort Augusta, dated Nov. 21, 1' American Archives, Vol. I, p. 1017.

owns on a mission of peace. They reported un-avorably.* The plan of the campaign was to form junction before reaching the Indian villages, and ewis accordingly halted at the mouth of the Ken-

* On the 14th day of October, 1774, an express arrived at Villiamsburg, Va., with a talk between Dunmore and chiefs of the ix Nations and Delawares, and also with some of the Mohawks, he time and place not named; probably at Pittsburg or its vi-inity.

Part of Lord Dunmore's reply to the Delaware and Six Nations' hiefs.—"Brethren: I am much obliged to you for the pains you ave taken to heal the sores made by the Shawanese. I would ave been very glad to have now given you a more favorable an-wer as to them; but you yourselves must be well acquainted how ttle the Shawanese deserve the treatment or appellation of rothers from me, when in the first place they have not complied ith the terms prescribed to them by Colonel Bouqet (and to hich they assented), of giving up the white prisoners; nor have ey ever truly buried the hatchet, for the next summer after that reaty they killed a man upon the frontiers of my government. he next year they killed eight of my people upon Cumberland iver, and brought their horses to their towns, where they dis-osed of them (together with a considerable quantity of peltry) the traders from Pennsylvania. Some time after, one Martin, trader from my country, was killed, with two men, on the Hock-ocking by the Shawanese, only because they were Virginians, at he same time permitting one Ellis to pass, only as he was a Penn-ylvanian. In the year 1771 twenty of my people were robbed by hem, when they carried away nineteen horses, and as many wned by Indians, with their guns, clothes, etc., which they deli-ered up to one Callender and Sprague, and other Pennsylvania raders, in their towns. In the same year, on the Great Kenhawa, my government, they killed ***, one of my people, and wound-d his brother; and the year following, Adam Strand, another f my people, his wife, and seven children, were most cruelly mur-ered on Elk waters. In the next year they killed Richards, an-ther of my people, on the Kenhawa. A few moons after they illed Russel, one of my people, and five white men and two ne-roes, near Cumberland Gap, and also carried their horses and effects to their towns, where they were purchased by the Pennsylvania raders. All these, with many other murders, they have commit-ed upon my people before a drop of Shawanese blood was spilt y them, and have perpetrated robberies upon my defenceless rontier inhabitants, which, at length, irritated them so far that

hawa, on the 6th of October, for communication an
orders from the Commander-in-chief. While ther
he encamped on the ground now occupied by th
village of Point Pleasant, without intrenchments o

they began to retaliate. I have now stated the dispute betwee them, and leave it to you to judge what they want."

Captain Pike, who had been on a mission to the Shawanese said:—"At my arrival at the lower towns I was told by Cornstal that he was much rejoiced to hear from his brothers, the whit people, in the spring, upon the first disturbances; that he had i consequence ordered all his young people to remain quiet, and no to molest the traders, but to convey them safe to their granc fathers, the Delawares, where they would be safe. The Shawa nese chiefs declared they were well pleased to hear from thei brothers, the English, and that they had spoken to all their youn people to remain quiet. Upon my arriving at the Standing Ston (now Lancaster, Ohio), I sent word to the Shawanese to assembl their counsellors; but as they were out hunting it could not b immediately effected. The principal warriors listened to th chiefs, and had no hostile intentions. The mischief which ha been done was perpetrated by the foolish young people; but tha now, as soon as they were assembled, they would be able to pre vent any thing of that nature for the future."

The Shawanese said:—"That a party of Twightwees and o Tawas, and a party of Wyandots, were as far advanced on thei way to war against the white people, as their towns, but that the had advised them to return; that they expected that the wa which threatened them would be extinguished, as they now desire peace."

The Mohegans delivered the following speech to the Shawa nese:—"Brethren: Formerly you came to us on the other sid of the mountains, and told us we were your elder brothers, desi ing us to come over and show ourselves to your grandfathers, th Delawares, that they might know our relationship. We did so and our people held fast the same chain of friendship; but no we see you only holding with one hand, while you keep a toma hawk in the other. We desire you, therefore, to sit down, an not be so haughty, but pity your women and children. We there fore take the tomahawk out of your hands, and put it into th hands of your grandfathers, the Delawares; they are good judges and know how to dispose of it."

The Shawanese replied:—"Brethren: We are glad to hea what you have said, and that you have taken the tomahawk ou of our hands, and given it to our grandfathers, the Delaware other defences. On the morning of the 10th of October he was attacked by one thousand chosen warriors of the western confederacy, who had abandoned their towns on the Pickaway plains to meet the Virginia troops, and give them battle before the two corps could be united. The Virginia riflemen occupied a triangular point of land, between the right bank of the Kenhawa and the left bank of the Ohio, accessible only from the rear. The assault was there-

But, for our part, we are not sensible that we have had the tomahawk in our hands. It is true some foolish young people may have found one out of our sight hid in the grass, and may have made use of it. But that tomahawk which we formerly held has been long since buried, and we have not since raised it. I heard some of the young people express a threat at the Delawares for interfering so much with their quarrel with the white people; that if they had any thing to say they wondered why the white people did not come themselves to speak."

Captain White Eyes, in behalf of the Delawares, to Lord Dunmore:—"Brother: As your brothers, the Shawanese, are desirous to speak to you by themselves, I hope you will listen to them. I will desire them to speak to you, and would be glad to acquaint them when they could see you to enter into a conference."

The Big Appletree, a Mohawk, said:—"This day it hath pleased God that we should meet together who are sent on behalf of another nation. The Shawanese told me that they heard there was something yet good in the heart of the Big Knife. They desired me to take their hearts in my hands, and speak strongly in their behalf to the Big Knife. I am glad the Shawanese, my younger brother, have desired me to undertake this business, and am equally rejoiced at the appearance thereof from your good speeches."

His address in answer to the reply:—"Brothers: I have already informed you of the evil disposition of the Shawanese towards us; but to convince you how ready the Big Knife is to do justice at all times, even to their greatest enemies, at the request of my brothers the Six Nations, and you the Delawares, I will be ready and willing to hear any good speeches which the Shawanese may have to deliver to me, either at Wheeling, where I now propose to be, or, if they would not meet me there, at the Little Kenhawa, or somewhere lower down the river."—American Archives, 4th Series, Vol. I, pp. 873—876.

fore in this quarter. Within an hour after the scouts had reported the presence of the Indians, a general engagement took place, extending from one bank of one river to the other, half a mile from the point.

Colonel Andrew Lewis, who seems to have been possessed of military talent, acted with steadiness and decision in this emergency. He arranged his forces promptly, and advanced to meet the enemy in force equal to his own. Colonel Charles Lewis, with 300 men, forming the right of the line, met the Indians at sunrise, and sustained the first attack. Here he was mortally wounded in the onset, and his troops, receiving almost the entire weight of the charge, were broken and gave way. Colonel Fleming, with a portion of the command, had advanced along the shore of the Ohio, and in a few moments fell in with the right of the Indian line, which rested on the river.

The effect of the first shock was to stagger the left wing as it had done the right, and its commander also was severely wounded at an early stage of the conflict, but his men succeeded in reaching a piece of timber land, and maintained their position until the reserve, under Colonel Field, reached the ground. It will be seen, by examining Lewis' plan of the engagement, and the ground on which it was fought, that an advance on his part and a retreat of his opponent necessarily weakened their line, by constantly increasing its length, if it extended from river to river, and would eventually force him to break it or leave his flanks unprotected. Those acquainted with Indian tactics inform us that it is the great point of his generalship to preserve his flanks and overreach those of his enemy. They continued, therefore, contrary to their usual practice, to dispute the ground with the pertinacity of veterans along the whole line, retreating slowly from tree to tree till one o'clock, P. M., when they reached a strong position. Here both parties rested, within rifle range of each other, and continued a desultory fire along a front of a mile and a quarter until after sunset.

The desperate nature of this fight may be inferred from the deep-seated animosity of both parties towards each other, the high courage which both possessed, and the consequences which hung upon the issue. The Virginians lost one-half their commissioned officers, and fifty-two men killed. Of the Indians twenty-one were left on the field, and the loss in killed and wounded is stated at 233.* During the night the Indians retreated, and were not pursued.

Having failed in this contest with the troops, while they were still divided in two parties, they changed their plan, and determined at once to save their towns from destruction by offers of peace.

Soon after the battle was over, a reinforcement of 300 Fincastle troops, and also an express from Lord Dunmore, arrived, with an order directing this division to advance towards the Shawanese villages without delay. Notwithstanding the order was given in ignorance of the engagement, and commanded them to enter the enemy's country unsupported, Colonel Lewis and his men were glad to comply with it, and thus complete the overthrow of the allied Indians.

The Virginians, made eager with success, and maddened by the loss of so many brave officers, dashed across the Ohio in pursuit of more victims, leaving a garrison at Point Pleasant. Our next information of them is, that a march of eighty miles, through an untrodden wilderness, has been performed,

* Burke's History of Virginia, Vol. III, pp. 392—8. American Archives, Vol. I, pp. 1015—1018, Letters of the Day.

and on the 24th of October they are encamped on the banks of Congo Creek (in Pickaway township, Pickaway county), within striking distance of the Indian towns. Their principal village was occupied by Shawanese, and stood upon the ground where the village of Westfall is now situated, on the west bank of the Scioto, and on the Ohio canal, near the south line of the same county. This was the head-quarters of the confederate tribes. In the mean time Lord Dunmore and his men had descended the Ohio to the mouth of the Great Hockhocking, established a depôt, and erected some defences called Fort Gower. From this point he probably started the express directed to Lewis at the mouth of Kenhawa, about fifty miles below, and immediately commenced his march up the Hockhocking into the Indian country; for the next that is known of him he is in the vicinity of Camp Charlotte, on the left bank of Sippo Creek, about seven miles south-east of Circleville, where he arrived before Lewis reached the station on Congo, as above stated. Camp Charlotte was situated about four and one half miles north-east of Camp Lewis, on the farm now owned by Thomas J. Winship, Esq., and was consequently farther from the villages than the position occupied by the left wing. There has been much diversity of opinion and statement respecting the location of the town, and also in regard to the positions of Camp Charlotte and Camp Lewis. The associations connected with these places have given them an interest which will never decline. This is probably a sufficient excuse for presenting here in detail the evidence upon which the positions of these several points are established.

It was at the town that Logan delivered his famous speech. It was not made in council, for he refused to attend at Camp Charlotte, where the talk vas held, and Dunmore sent a trader, by the name of John Gibson, to inquire the cause of his absence. The Indians, as before intimated, had made propositions to the Governor for peace, and probably before ie was aware of the result of the action at Kenhawa. When Gibson arrived at the village, Logan came to iim, and by his (Logan's) request, they went into an idjoining wood, and sat down. Here, after shedding ibundance of tears, the honored chief told his pathetic tory.* Gibson repeated it to the officers, who caused t to be published in the *Virginia Gazette* of that rear. Mr. Jefferson was charged with making improvements and alterations when he published it in iis Notes on Virginia; but from the concurrent testimony of Gibson, Lord Dunmore, and several others, t appears to be as close a representation of the original as could be obtained under the circumstances. The only versions of the speech that I have seen are iere contrasted in order to show that the substance and sentiments correspond, and that it must be the production of Logan or of John Gibson, the only white man who heard the original.

WILLIAMSBURGH, VA., Feb. 4, 1775.

The following is said to be a message from Captain Logan, an ndian warrior, to Gov. Dunmore, after the battle in which Colonel Charles Lewis was slain, delivered at the treaty: —

"I appeal to any white man o say that he ever entered Logan's cabin but I gave him meat, hat he ever came naked but I lothed him.

NEW YORK, Feb. 16, 1775.

Extract of a letter from Virginia: —

"I make no doubt the following specimen of Indian eloquence and mistaken valor will please you, but you must make allowances for the unskilfulness of the interpreter:—

"'I appeal to any white man to say if ever he entered Logan's cabin hungry and I gave

* Affidavit of John Gibson, Jefferson's Notes, Appendix, p. 16.

"In the course of the last war Logan remained in his cabin an advocate for peace. I had such an affection for the white people, that I was pointed at by the rest of my nation. I should have ever lived with them had it not been for Colonel Cresap, who last year cut off, in cold blood, all the relations of Logan, not sparing my women and children. There runs not a drop of my blood in the veins of any human creature. This called upon me for revenge. I have sought it. I have killed many, and fully glutted my revenge. I am glad there is a prospect of peace on account of the nation; but I beg you will not entertain a thought that any thing I have said proceeds from fear. Logan disdains the thought. He will not turn on his heel to save his life. Who is there to mourn for Logan? Not one."

him not meat; if ever he came cold or naked and I gave him not clothing.

"'During the course of the last long and bloody war Logan remained in his tent an advocate for peace. Nay, such was my love for the whites, that those of my own country pointed at me as they passed by, and said, —'Logan is the friend of white man.' I had even thought to live with you but for the injuries of one man. Colonel Cresap the last spring, in cool blood and unprovoked, cut off all the relatives of Logan, not sparing even my women and children. There runs not a drop of my blood in the veins of any human creature. This called on me for revenge. I have sought it. I have killed many. I have fully glutted my vengeance. For my country I rejoice at the beams of peace; yet do not harbor the thought that mine is the joy of fear. Logan never felt fear. He will not turn on his heel to save his life. Who is there to mourn for Logan? Not one.'"

The right hand translation is literally the same as the copy given in Jefferson's Notes, page 124, and is doubtless the version given out by himself at the time. The authenticity of the ideas, and, if not the words, at all events the style, is in some degree sustained by another piece of Logan's composition, which was found tied to a war-club at the house of one Robertson, in Fincastle county, Va., after a massacre of his family by the Indians. Logan had previously caused it to be written with a burnt stick, by a prisoner named William Robinson, saying he would kill somebody, and leave the letter in the house.

"Captain Cresap, what did you kill my people on Yellow ek for? The white people killed my kin at Conestoga a great le ago, and I thought nothing of that; but you killed my kin in at Yellow Creek, and took my cousin prisoner. I thought ust kill too. I have been three times to war since; but the ians are not angry, only myself.

"July 21, 1774. CAPT. JOHN LOGAN."

I have shown elsewhere that Logan was mistaken to the connection of Cresap with the murders at ker's.

It was repeated throughout the North American onies as a lesson of eloquence in the schools, and)ied upon the pages of literary journals in Great itain and the Continent. This brief effusion of ngled pride, courage, and sorrow, elevated the racter of the native American throughout the elligent world, and the place where it was delivered never be forgotten so long as touching eloquence dmired by men.

Camp Charlotte was situated on the south-west rter of section 12, town 10, range 21, upon a asant piece of ground, in view of the Pickaway ins. It was without permanent defences, or at st there are no remains of intrenchments, and is essiblo on all sides. The creek in front formed no pediment to an approach from that quarter, and country is level in the rear. Camp Lewis is said be upon more defensible ground on the north-east rter of section 30, same township and range. The encampments have often been confounded with h other.

The testimony which I shall here introduce was eloped in the year 1830, in a case in the Supreme urt of Ohio for Pickaway county, where questions ating to the position of Camp Charlotte and the illicothe towns were involved. It was furnished me by J. D. Calwell, Esq., of Chillicothe, counsel f defendants, in the suit of John Gibson's heirs v Duncan McArthur, and others. It consists of dep sitions taken upon the ground at Camp Charlotte, f the purpose of sustaining the description of a Virgin military warrant, which reads thus:—"August 2 1787—Entry No. 450. John Jolliff's (heir) ente 2,666⅔ acres; a military warrant No. 825, on t Sciota river, at the first fork above the old Chillicot town, which town is about seven miles from a pla called Camp Charlotte," etc.

Caleb Evans deposes and says that he was here thirty-t years ago, and noticed these stumps, and old John Hargus t me it was Camp Charlotte. Hargus said he was a captain spies with Dunmore, and when they came here the Indians crow ed them so that Dunmore and six or seven others, myself includ went across to meet Lewis, who was encamped on Congo (Cre on some knolls where Judge Barr now lives. We met Lewis j after he had left his camp to give battle to the Indians, and dered him back. I was here when there were no marks of an in these parts except at these two camps. Hargus and mys were the only persons then in this vicinity. The cattle of army were kept on a prairie across the creek.

Thomas Barr.—Deponent was here thirty-two years ago. went back to Pennsylvania and saw Colonel Williamson, who with Dunmore. We differed about the appearance of Camp Ch lotte, and after I saw this place the next season, and I went b again, we agreed in every particular. There was a spring and or six trees deadened by him in a drain. Lewis' camp is about miles west of this, on Congo. One Boggs came through the co try about that time, who was at the camp, and said this is the s Boggs told about Dunmore overtaking Lewis, and ordering hir stop, and not attack the Indian towns. I saw Mr. McIntyre, Zanesville, soon after that place was settled. Mr. McIntyre Mr. Zane were to lay out a road from Wheeling through Za ville, Standing Stone (now Lancaster), Chillicothe (meaning settlement in Ross county), to Limestone (now Maysville). T took an Indian Pilot to lead them to Chillicothe. He led then the place now called Westfall, and said that was Chillicothe, the only Chillicothe he knew of. This was about twenty-r years ago.

George Wolf.—I live near Camp Lewis, and came here thi

ree years ago. I heard John Hargus and John Boggs talk about is place, and call it Camp Charlotte.

James Moore.—I came to Chillicothe (in Ross county) in 1796. 1797 or 1798 I went upon the ground where Westfall now is. I oved there in 1799, and it appeared to have been an Indian town. he Indians who were about there said the inhabitants called emselves Chillicothe*es*, and their town Chillicoth*ee*. They had a wn on the north fork of Paint Creek of the same name, and anher on Mad River. They called this at Westfall Old Chillicoth*ee*. hese Indians were Shawanese, Delawares, and Wyandots, and said e reason the Chillicoth*ees* left was the prevalence of the small x. Describing the disease, the Indians said the people die—die -die—some day one, some day two; and they bury—bury—bury, d pointed out the graves.

Fergus Moore testified to same as James Moore, and dded that in digging the Ohio canal bones were und in great numbers at the places shown to them s graves.

This evidence will probably be considered as conlusive in relation to the position of Old Chillicothe nd the two camps.

Before Lord Dunmore reached the vicinity of the ndian towns he was met by a flag of truce, borne by white man named Elliot, desiring a halt on the part f the troops, and requesting for the chiefs an interreter with whom they could communicate.* To this is Lordship, who, according to the Virginians, had n aversion to fighting, readily assented. They furhermore charged him with the design of forming an lliance with the confederacy to assist Great Britain gainst the colonies in the crisis of the revolution, hich every one foresaw. He, however, moved forard to Camp Charlotte, which was established rather s a convenient council ground than as a place of ecurity or defence. The Virginia militia came here r the purpose of fighting, and their dissatisfaction

* Gibson's Affidavit.

and disappointment at the result amounted almost mutiny. Lewis refused to obey the order for a ha considering the enemy as already within his gras and of inferior numbers to his own. Dunmore, we have seen, went in person to enforce his orde and, it is said, drew his sword upon Colonel Lew threatening him with instant death if he persisted farther disobedience.

The troops were concentrated at Camp Charlott numbering about 2,500 men. The principal chie of the Scioto tribes had been assembled, and son days were spent in negotiations. A compact treaty was at length concluded and *four* hostag put in possession of the Governor to be taken to Vi ginia.* We know very little of the precise terms this treaty, nor even of the tribes who gave it the assent. It is said the Indians agreed to make t Ohio their boundary, and the whites stipulated not pass beyond that river.

An agreement was entered into for a talk at Pit burgh in the following spring, where a more full trea was to be made.†

At what precise time the British standard left t Pickaway plains, we are not informed, nor by wh route, after passing Fort Gower, or whether in a bo or in detachments, the troops made their way hon to the settlements. It is said that they reached Vi

* The following extract of a letter from Arthur St. Clair Governor Penn, dated Ligonier, Dec. 4, 1774, needs confirmati It is but one of many instances of the contradictory statemer which embarrass our conclusions in reference to the importa doings of the year 1774: —

"The Mingoes that live on the Scioto did not appear to tre and a party was sent to destroy their towns, which was effecte and there are (12) twelve of them prisoners in Fort Pitt."—Am ican Archives, Vol. I, p. 1013.

† American Archives, Vol. I, p. 1222.

;inia highly dissatisfied with the Governor and the reaty; but this dissatisfaction does not appear to ıave been general.

Dunmore had assumed the credit of the battle at 'oint Pleasant. The Virginians, who participated in hat action, denied that it was an event in which he ad the remotest concern; and not only was not ware of the affair till after it had occurred, but had either anticipated or desired its occurrence.* The roops were not paid, and they represented the whole roceeding as a method of forming an alliance with he western confederacy, of which fighting formed no art in his Lordship's plan. His position was one of ifficulty, and he seems to have been deficient in the ualities of prudence, determination, and self-comıand, so necessary to one thus situated. He is epresented as a haughty, wayward, and unapproachble person, with a selfish, hesitating, and overbearıg mind.

In addition to the scattered items of this expedi.on here given, I will add a statement which comes ery well authenticated, but seems to contradict other 'ell known facts. It is in relation to another camaign to the Indian country by Dunmore in the year ɔllowing, or 1775. It was related to me by Walter 'urtis, Esq., of Belpre, Washington county, Ohio, nd I think transmitted by him in substance to the ecretary of this Society. Mr. Curtis received it om General Clark, an eminent citizen of Missouri, brother of General George Rogers Clark, of Kenıcky.

In 1831 a steamboat was detained a few hours ear the house of Mr. Curtiss, on the Ohio, a short istance above the mouth of the Hockhocking, and

* Burke's History of Virginia, Vol. III, p. 406.

General Clark came ashore. He inquired respectin the remains of a fort or encampment at the mouth o the Hockhocking river, as it is now called. He wa told that there was evidence of a clearing of severa acres in extent, and that pieces of guns and musket had been found on the spot; and also that a collec tion of several hundred bullets had been discovered o the bank of the Hockhocking, about twenty-five mile up the river. General Clark then stated that th ground had been occupied as a camp by Lord Dur more, who came down the Kenhawa with 300 men i the spring of 1775, with the expectation of treatin with the Indians here. The chiefs not making the appearance, the march was continued up the rive twenty-five or thirty miles, where an express fro Virginia overtook the party. That evening a counc was held, and lasted till very late at night. In th morning the troops were disbanded, and immediatel requested to enlist in the British service for a state period. The contents of the dispatches had not tran pired when this proposition was made. A major c militia, by the name of McCarty, made an harang to the men against enlisting, which seems to ha been done in an eloquent and effectual manner. H referred to the condition of the public mind in th colonies, and the probability of a revolution, whi must soon arrive. He represented the suspicious ci cumstances of the express, which was still a secret the troops, and that appearances justified the co clusion that they were required to enlist in a servi against their own countrymen, their own kindre their own homes. The consequence was that but fe of the men re-enlisted, and the majority, choosi the orator as a leader, made the best of their way Wheeling. The news brought out by the couri proved to be an account of the opening combat of t

Revolution at Lexington, Mass., April 20, 1775. General Clark stated that himself (or his brother) was in the expedition.

Lord Dunmore is said to have returned to Virginia by way of the Kenhawa river.

There are very few historical details sustained by better authority than the above relation. Desirous of reconciling this statement with history, I addressed a letter to General Clark, requesting an explanation, but his death, which happened soon after, prevented a reply. It would be as difficult to pronounce it an entire error as to give it full belief.

On the 20th of April, Dunmore had lost all influence in Virginia, entrenched himself in his house at Williamsburg, and removed the powder from the magazine on board the Fowey, a British vessel of war. The people were then in arms, not for the purpose of organizing war parties against the Indian country, but intending to assault the troops and marines of England, which the Governor had posted on his premises to ensure the safety of, or to prevent access to, his person. Before the 17th of July he had abandoned the capital, and removed to the Fowey with his family and papers. There is no mention of more than one expedition in the history of Virginia, and he is stated to have been there when the battle of Lexington was first known, and is accounted for from that time until August of the subsequent year.

I leave it for further information to refute or establish the truth of this narrative, and offer it here because no shadow of evidence respecting the transactions of that interesting period in the west ought to be neglected. For the same reason I attach the following copies of resolutions, etc., taken from the "American Archives," a most valuable publication, issued and issuing at the expense of the United States.

In this publication the government has spared nc pains to obtain correspondence in all the States, and in foreign countries, illustrating our history prior to the close of the Revolution. The first extract has some relation to the statement of General Clark of a meeting of the officers under Dunmore, but in a different year.

"At a meeting of the officers under the command of His Excellency Rt. Hon. Earl of Dunmore, convened at *Fort Gower*, Nov. 5, 1774, for the purposes of considering the grievances of *British America*, an officer present addressed the meeting in the following words:—

"'Gentlemen: Having now concluded the campaign, by the assistance of Providence, with honor and advantage to the colony and ourselves, it only remains that we should give our country the stronger assurance that we are ready at all times, to the utmost of our power, to maintain and defend her just rights and privileges. We have lived about three months in the woods, without any intelligence from Boston, or from the Delegates at Philadelphia. It is possible, from the groundless reports of designing men, that our countrymen may be jealous of the use such a body would make of arms in their hands at this critical juncture. That we are a respectable body is certain, when it is considered that we can live weeks without bread or salt,—that we can sleep in the open air without any covering but that of the canopy of heaven,—and that we can march and shoot with any in the known world. Blessed with these talents, let us solemnly engage to one another, and our country in particular, that we will use them for no purpose but for the honor and advantage of America, and of Virginia in particular. It behoves us, then, for the satisfaction of our country, that we should give them our real sentiments by way of resolves, at this very alarming crisis.'

"Whereupon the meeting made choice of a committee to draw up and prepare resolves for their consideration, who immediately withdrew, and after some time spent therein, reported that they had agreed to and prepared the following resolves, which were read, maturely considered, and agreed to nem. con. by the meeting, and ordered to be published in the Virginia Gazette:—

"'Resolved, That we will bear the most faithful allegiance to His Majesty King George the Third, whilst His Majesty delights to reign over a brave and a free people; that we will, at the expense of life and every thing dear and valuable, exert ourselves in the support of the honor of his crown and the dignity of the British Empire; but as the love of liberty and attachment to the real interests and just rights of America outweigh every other consideration, we resolve that we will exert every power within us for the defence of American liberty, and for the support of her just rights and privileges, not in any precipitous, riotous, or tumultuous manner, but when regularly called forth by the unanimous voice of our countrymen.

"'Resolved, That we entertain the greatest respect for His Excellency the Rt. Hon. Lord Dunmore, who commanded the expedition against the *Shawanese*, and who, we are confident, underwent the great fatigue of this singular campaign from no other motive than the true interests of the country.'

"Signed, by order and in behalf of the whole corps,

"BENJAMIN ASHBY, Clerk."

These resolutions bear date only ten days after the arrival of Lewis at Camp Charlotte. Of this time at least four days must have been occupied in the march, which must have exceeded eighty miles in distance, and we may infer that the troops moved from the Indian towns about the 1st of November, 1774.

We are not able to determine whether Lord Dunmore was present when these resolutions were adopted. On the 12th instant he is found at fort Burd, near Pittsburg, sitting in judgment upon one of the refractory Pennsylvanians, for violating the Virginia proclamation. He arrived at Williamsburg, Va., on the 4th of December, and received the attentions and congratulations of the public authorities. (American Archives, Vol. I, p. 1018.)

It is highly probable that the army was disbanded at Fort Gower, and came home in different parties, and by such routes as were nearest and most convenient. The hostages had not arrived at Williamsburgh at the above date; in fact (12) twelve of them were left at Fort Dunmore, as the Virginians called Fort Pitt.

The delivery of white prisoners and horses in possession of the Indians appears to have taken place at Point Pleasant early in February.

"WILLIAMSBURG, Va., Feb. 10, 1775.

"A private letter from the frontiers gives an account that the Cornstalk King of the *Shawanese* nation, a few days ago, arrived at the mouth of the Great Kenhawa, where Captain Russell is stationed, and delivered to him several of the old white prisoners, and a number of horses, agreeable to Lord Dunmore's desire. The Cornstalk informs that every thing at present is peaceable and quiet in the quarter he left, but that he would not undertake to say how long that pacific disposition would last, as the Pennsylvanians have sent some of their traders there, who were endeavoring all they could to persuade them that Lord Dunmore's view, in bringing the hostages to Williamsburg, was to deceive them, and that whenever it was in his power to raise another army he would immediately take every advantage and cut them off. This kind of reasoning had no material effect, it seems, as the Indians throughout the different tribes entertain the highest opinion of his Lordship's conduct with respect to his late manœuvres on the frontiers.

"This morning we received information from a gentleman at the Ohio that the Mingo Indians have killed three of the Delawares, which gives much concern to the neighboring white people. The Pennsylvanians, it appears, are greatly blamed, as they use every artifice in their power to create discontent and jealousy among the Indians. Our correspondent says they took one of our constables and immediately confined him in one of their jails, upon which *two* companies of the Virginians assembled, being determined to rescue him, which they did, together with some others which they served in the same manner, and also pulled down the jail. The Mingoes, we are likewise informed, are very desirous to see Lord Dunmore, in order fully to comply with his terms, and to make a lasting peace with him."—American Archives, p. 1226.

ANTIQUITIES OF AMERICA.*

[Hesperian, July, 1839.]

THAT a work full of learned research, executed in a pure and pleasing style of language, abounding in college lore, arranged with logical accuracy, and expressed with argumentative force, should make its appearance *at the West*, excites extreme wonder among the salt-water literati. Hear the North American:—"A quarto volume, from what, when we studied geography, used to be known by the instructive name of the 'territory north-west of the Ohio,' is something to attract attention. And when we open it, and find it printed in a style which emulates the London press, and is seldom even attempted in America, we turn to the title-page again, to see if we did not mistake its birth-place. That one of the community in that great pork-mart (Cincinnati) should write a work upon a subject requiring long study and deep thought, is to us a pleasing fact."

Heretofore speculations relative to the objects and origin of our ancient works have made their appearance at the east, and far from the interesting

* Notice of an Inquiry into the origin of the Antiquities of America. By John Delafield, Jr. With an Appendix, containing notes, and a "View of the Causes of the Superiority of the Men of the Northern over those of the Southern Hemisphere." By James Lakey, M. D. Cincinnati: published by N. G. Burgess & Co. Stereotyped by Glezen & Shepard. 1839.

remains so profusely discussed. The American Antiquarian Society at Worcester, Massachusetts, made the first attempt at a regular exploration, by employing Mr. Atwater to survey and describe them in 1819. The first volume of their Archælogea Americana contains the results of this gentleman's labors; and, considering the time of its appearance, and the means appropriated to the design, much information was thus spread before the world. The work, however, contains but a portion of these ruins, and some of the most remarkable are still undescribed. Subsequent examinations have, moreover, thrown discredit upon some of the representations made in the Archælogea, and in points upon which theories have been erected, both by the author and others.

About 1833 Mr. Joseph Priest, of Albany, issued the third edition of a book composed apparently from the relations of travellers, or the publications of Mr. Atwater and the Rev. Mr. Harris. The last-named gentleman came very early to Ohio, and located with the Ohio company at Marietta. In 1803 his "Tour" was published, and is worthy of credit. But the production of Mr. Priest, though highly amusing as a collection of wonders, will rank more properly with the tales of the "Tongo Islands," as a work of authority. He has, fortunately, enlightened us in some cases as to the source from whence these fictions were derived. An Englishman, by the name of Ash, has palmed upon him at least two entire descriptions of his own manufacture. We refer to the grave near Marietta, with mats and hieroglyphics, and the cavity near Zanesville, containing metallic spheres. So far as we know there does not exist such a description or descriptions as will convey to non-residents a proper and full idea of the ancient works that remain among us.* And, consequently, the numerous speculations hazarded abroad in regard to their design, antiquity, and present appearance, rest upon false, or at least, uncertain premises. How proper it is, then, that those who discuss the subject should be eye witnesses. And, aside from the assistance thus gained to truth, who can enter upon the investigation with the ardor of one standing upon the tumulus itself, the sacred altar of that by-gone race, whose origin is so deeply obscured by the mists of unrecorded ages.

To wander along lines of embankment, thrown up in every variety of form and dimension, parallels and squares, circles and ellipess, and every combination of curve and right line, is not the gratification of a mere idle curiosity. The observer catches an inspiration from the associations of the place. There, in the solitude of the forest, lie the uudoubted works of human hands; but, by whom erected? When? For what purpose? What language once sounded through the air? What feats of war, devotions of religion, acts of wisdom, or deeds of cruelty were enacted here? All is unknown; wrapt in inscrutable mystery; not a line carved, nor a record left, nor even a tradition transmitted whereby we can form a satisfactory conclusion. A strong enthusiasm comes upon the mind, and every step along the ditches, over the mounds, or down the excavations, raises the intensity of interest awakened by such a presence.

Mr. Delafield, acting under the full weight of those exciting mysteries, and feeling all the ambition natural to an inquisitive observer, to work out a solution where the world was lost in wonder, applied himself unremittingly to an examination of the authors upon

* This deficiency is now supplied by the publication in 1848, of Vol. I, of the Smithsonian Contributions to Knowledge.

the antiquities of Asia and South America. We know not whether the idea that the race of the mounds might be identical with the Caucasian race was original with him or not. It was a bold thought, and though apparently a wild one, he has led us from fact to fact, and deduction to deduction, till we are more surprised at the clearness of the proof than the grandeur of the conception. And whatever may be the truth in regard to the originality of this doctrine, when we consider the nature and obscurity of the case, there can be but one opinion as to the merits of the researches of the author, gathering from the four quarters of the globe corroborative testimony of astonishing force. The work may be considered as an abstract of the heretofore scattered facts, bearing upon that question. In addition to a most judicious selection of evidence, Mr. D. has fortunately obtained some striking auxiliary facts not before public, and which fell within his reach as a resident of the country where the works are found.

Blumenbach divides the human race into three families, because he found three marked classes of crania, and refers the origin of our race to the Eastern Continent. In viewing the head or skull from above, looking downwards, a method of comparison called *norma verticalis*, he could arrange all crania in three parcels, from a similarity of outline or horizontal projection. The original families are called the Caucasian, of Southern Asia, the Mongolian, of Northern Asia, and the Ethiopian. Whether this anatomical distinction is traced to the three sons of Noah, as the head of each grand division of the human family, we are unable to state.

It has been remarked above, that the book under consideration is written to sustain the theory that the *race of the mounds came from Asia.* The first proposition advanced is this: The Peruvians came from Mexico. Second: The Mexicans were from the North. Opposite page seventeen, is a lithographic representation of an ancient Peruvian skull, taken from the temple of the Sun; and in the same plate, two crania, obtained near Bogota, with a fourth taken from a mound in Cincinnati. The coincidence of form in these heads goes to sustain both propositions. Vega, book 3, chap. 7, says: The Peruvians built bridges of withes. Clavigero, Vol. I, p. 389, says, the Mexicans did the same thing. Ulloa, who spent ten years in Peru, Mexico, and Colombia, says: "If we have seen one American, we have seen all, their color and make are so nearly alike.' Chronica Del Peru, part 1, chap 19. Copan, a country between Mexico and Peru, was settled by Toltecas, an ancient Mexican tribe. Letter of J. Gulindo, Archælogea Americana, Vol. II. When the Spaniards gained a footing in South America they found the mountain ranges of the Cordilleras "were the abodes of a high state of civilization—the residences of nations dwelling in cities, skilful in the texture of cloths, ingenious in the mechanical arts, and possessing no small acquaintance with astronomy and general science." "Among these people have been found *national annals and records* which go back to a period corresponding with our *sixth century*, and relate the name of the illustrious emperor Citin, who led from the unknown regions of Azatlan and Teocolhuacan, the *northern nations* into the plains of Anahuac," p. 15. Azatlan means "near water," and Teocolhuacan "in the midst of the houses of God." The comparison of crania also establishes a plain difference between the present North American Indian and the race of the mounds. The same examination farther gives an identity of the North American Indian, with the Mongolian, or Tartar race, showing a successive emigration of the two Asiatic families to the American Continent. But we allow the author to state his own case.

As this essay is a chain of facts, collected from many authors, and each forms a link in the concatenation, the loss of one of which may break at once the argument to be deduced, it were well to state the position we now occupy, viz: That we have traced the descendants of that race which constructed our ancient works, by the following train of argument:

I. The extension of tumuli, &c., through Western North America and Mexico to Peru, induces a belief that the race which constructed the memigrated thither; and their termination there leads to the conclusion that the nation went no farther.

II. The traditions of the North American Indians assert distinctly their ejectment of a people from the present region of Western North America, who correspond with the native Mexicans, and who emigrated hence.

III. On the discovery of America, a tract of country occupying the present limits of Mexico, Colombia, and Peru, was in a high state of civilization, while all around them was shrouded in mental darkness.

IV. National annals have been found among the Mexicans, expressly stating that a period corresponding to our sixth century, their ancestors emigrated from the north, under the guidance of their illustrious Emperor, Citin, or Votan.

V. Traditions assert that the introduction of civilization into Peru was by the emigration of certain wise men from Mexico.

VI. Anatomical research exhibits a striking coincidence between the crania of the race of the mounds, and of the ancient Peruvians, differing from all others in the world, and proving conclusively that they were a distinct race from the ancestors of our present Indian tribes.

We propose now an investigation of the inquiry,

"whence is this family descended, and where were their ancient homes?"

In pursuing systematically the chain of evidence, it is proposed to divide the argument into the following branches:

1. The evidence from comparative philology.
2. That drawn from anatomy.
3. That deduced from their mythology.
4. That arising from their hieroglyphical writings.
5. That drawn from their astronomy.
6. The evidence derived from their architecture and decorations.
7. That deduced from their manners and customs.

To give the proof or the arguments belonging to the several heads of discussion would be a compilation of the book. It is already so much compressed as not to admit of an abstract.

The sub-division called "Philological Evidence," traces first the resemblance in orthography, between words having the same meaning in the Asiatic and North American tongues.

There are eighteen words which have a most perfect resemblance.

The inquiry may be made, "What number of words, found to resemble one another in different languages, will warrant our concluding them to be of common origin?" The learned Dr. Young applied to this subject the mathematical test of the calculus of probabilities, and says, "it would appear therefrom that nothing whatever could be inferred with respect to the relation of any two languages, from the coincidence of sense of any single word in both of them; the odds would be three to one against the agreement of any two words; but if three words appear to be identical, it would be then more than ten to one that they must be derived in both cases from some parent language, or introduced in some other manner; six words would give more than seventeen hundred chances to one; and eight, near one hundred thousand: so that in these cases the evidence would be little short of absolute certainty."*

Applying the same method of calculation to the terms used by the southern Asiatics, and the South Americans, a considerable list is found to be common to both. "Cami" is the term for the god in Japan; "Cemi" that of the deities of Mexico. In Sanscrit "indre" is the sun, "manya" love, "vipulo" great; in the language of the Incas of Peru, "inti" the sun, "munay" love, "veypul" great.

The next division relates to anatomy, and here the connection of the Northern American, and the Northern Asiatic, is first introduced.

"The portrait painter, Mr. Smibert, who accompanied Dr. Berkeley, then Dean of Derry, afterwards Bishop of Cloyne, from Italy to America, in 1728, was employed by the Grand Duke of Florence to paint two or three Siberian Tartars, presented to the Duke by the Czar of Russia. Mr. Smibert, on his landing at Narragansett Bay with Dr. Berkeley, instantly recognized the Indians to be the same people as the Siberian Tartars whose pictures he had taken. I shall show that the language of the Siberian Tartars and that of the Tongousi have an extensive range in North America."†

The Mongolian race, as the American, contains several sub-divisions, many tribes possessing dissimilar customs, habits, and languages. But throughout the whole north of Asia we find this family leading a nomadic or roving and savage life. Equally given to war and to the chase, they both reject the light of civilization gleaming over their southern borders.

Illustrative of this branch, a lithograph of the cranium of an Egyptian Mummy is given, and con-

* Philosophical Transactions, CIX, for 1818, p. 70.
† Dr. S. B. Barton, pp. XVI, XVII.

trasted with another, from an ancient burying ground near Lima. The sketches are taken from originals now in Cincinnati, and are doubtless correct. The resemblance, however, is not strong enough to give great support to the anatomical argument.

We come now to the "Mythological Evidence."

Here, again, is a coincidence between the aborigines of America and the southern Asiatics, that we cannot fairly attribute to mere chance.

"The Mexicans had some ideas of a *supreme* God, to whom they gave fear and adoration. They did not represent him by any visible form, calling him 'Teotl,' or God, to whom they applied expressions highly characteristic of his nature. They also believed in an evil spirit, called '*Tlacatecolotl*,' or 'rational owl.'" *

This quotation bears more directly upon the connection between the race of the mounds and the Mexicans. It is part of a statement showing a belief in the metempsychosis, or transmigration of souls, common to the Hindoo and the Mexican, that follows the description of a painting which is copied in the work.

The opposite plate is the copy of a Mexican painting taken from the Codex Vaticanus, at Rome, whither it arrived from the new continent, shortly after the early conquests in New Spain. It will be found in the Paris folio edition of

* The Mexicans were in the habit of worshipping rude sculptures of this evil spirit, to prevent his anger, and consequent dangerous power. One of these images was dug out of a large tumulus in the city of Columbus, the capital of Ohio, and was exhibited to the Historical Society when an abstract of this essay was read by the author. It is an owl rudely carved out of a block of sand-stone, on the back of which are two holes apparently bored by a conical instrument, and in such a direction as to meet at the points, so that a thong can be passed through by which the idol can be suspended."

Baron Humboldt's "Vues des Cordilleres." The large figure represents the celebrated "serpent woman," Cihuacohuatl, called, also, Tonacacihua, "woman of our flesh." The Mexicans considered her the mother of the human race. She is always represented with a great serpent; but for this no reason is assigned, as though, in process of time, part of the tradition were lost. Behind the serpent, who appears to be speaking to Eve, are two naked figures, of different color, and in the attitude of contention. The serpent woman was considered at Mexico as the mother of twin children, and which are here represented. This part of the painting is entirely unexplained. Baron Humboldt supposes they represent Cain and Abel, of Semitic tradition He considers the other figures, however, merely as vases, respecting which a quarrel may have ensued. I would respectfully suggest that (if so much be conceded, as is necessarily true, that the chief figures are Eve, the serpent, Cain and Abel) then the others are the two altars, one of which, standing erect, bears the offering of Abel, viz: a ram, the horns of which are rudely delineated; while the other is the altar of Cain, rejected by the Almighty, and therefore painted upside down, containing his offering, viz: the fruits of the earth. Baron Humboldt thinks the difference of the color of Cain attributable perhaps to fancy or chance. May we not consider it typical of the mark set on the murderer by Jehovah for the heinousness of his guilt? For it will be noticed that Abel is represented with the same tint as Eve; and from the general care in the distribution of colors through the piece, we can not infer want of design.

A tradition exists among the native Mexicans bearing close analogy to the Semitic account of the flood, the building of the tower of Babel, and its destruction; and which corresponds with the early traditions of Xisthurus of the Hindoos.

One or two copious extracts from this division of the subject seems to be necessary.

The following description of the Mexican cosmogony is

condensed from the valuable work of Baron Humboldt, "des anciens monumens de l'Amerique."

The sacred books of the Hindoos, especially the Bhagavita Pourana, speak of the *four ages*, and of the pralayas, or cataclysms, which at different epochs have destroyed the human race. Gomara, in his *Conquista, fol.* 119, says that the natives of Culhua, believe according to their hieroglyphical paintings, that, previous to the sun which now enlightens them, four had already been extinguished. These *four suns* are as many ages, in which our species has been annihilated by inundations, by earthquakes, by a general conflagration, and by the effect of destroying tempests. The Codex Vaticanus, at Rome, No. 3738, contains the drawings which are represented on the annexed pages, being copies of native hieroglyphic paintings, made by the Dominican monk, Pedro de los Rios, A. D., 1566. They illustrate the destruction of the world at the expiration of each age, and are described in a very curious history, written in the Aztec tongue, fragments of which have been preserved by the native Mexican, Fernando de Alvar Ixtlilochitl. The testimony of a native writer, and the copies of Mexican paintings, made on the spot, merit, undoubtedly, more confidence than the recital of the Spanish historians.

The four hieroglyphical paintings are given in full. The four cycles, or ages, are four thousand to five thousand years each, called the "age of Justice" (five thousand two hundred and six years); "age of fire," (four thousand eight hundred and four years); "age of wind," (four thousand and ten); "the age of the flood," (four thousand and eight.) A man and a woman escapes from each cataclysm, indicating a coincidence with the Jewish Scriptures, and also with the Hindoo belief.

Hieroglyphics.—Our knowledge of hieroglyphical writing is confined to the Egyptian productions, of late years so fully elucidated by the Champollions. There are three kinds or degrees of this method of making records: the *phonetic*, *figurative*, and *symbol-*

ical. The first expresses sound like an alphabet, an constitutes much of the Egyptian writing. Cham pollion read the names of Ptolemy and Cleopatra, o the Rosetta monument, by a resort to this methoc Humboldt says (vues des Cordilleres, pp. 64–5 "There are, in Mexico, remains of those hieroglypl ics, called phonetic, having relation not to the thing but to the spoken name. The phonetic system of th Toltecans is intelligible at first glance. The head o a Toltecan king appears, along with others, in th pyramidal tower of Palenque." The name is inscribe over it in a rectangle or cartouche, after the Egyp tian fashion, and reads Acatla Potzin. The secon method, the *figurative*, "was in common use amon the Mexicans, and forms no small portion of the: scriptural remains," p. 45. As to the third, or *syn bolical*, "the Mexicans not only represented the sin ple images of objects, but they had some character answering, like the signs of algebraists, for thing devoid of figure, or difficult of representation," p. 4

Astronomical Evidence.—"The civil year of th Aztecs is a solar year of three hundred and sixty five days. It was divided into eighteen months (twenty days, making three hundred and sixty days to which they added five days, and began the yea anew.

The Peruvian year was divided, as is customary in south ern Asia, into twelve moons [guilla], the synodical revolu tions of which end at three hundred and fifty-four day, eight hours and forty-eight minutes. To correct the luna year, and make it agree with the solar, they added, accor ing to an ancient custom, eleven days, which, after an edi from the Incas, were distributed among the twelve moon

But perhaps a still more striking instance presents itse to us in a comparison of the zodiacal signs of southern Asi and this civilized Aboriginal race of America. Barc [umboldt collected and arranged in a tabular form the ames, of the Mexican hieroglyphic zodiacal signs. They ere compiled by him from the various writers of the six- enth century. From this it appears that a great propor- on of the names by which the Mexicans indicated the venty days of their month, are those of a Zodiac used nce the remotest antiquity by the inhabitants of eastern sia.

These quotations we consider very positive evidence of ı early identity between the aboriginal race of America, ıd the southern Asiatic and Egyptian family. To con- ude the testimony on this point, the following extract of a tter of Mr. Jomard is adduced:

"I have also recognized, in your memoir on the division ' time among the Mexican nations compared with those of sia, some very striking analogies between the Toltec char- ters and institutions observed on the banks of the Nile. mong these analogies there is one which is worthy of at- ntion. It is the use of the vague year of three hundred ıd sixty-five days, composed of equal months, and of five mplementary days, equally employed at Thebes and [exico, a distance of three thousand leagues. It is true ıat the Egyptians had no intercalation, while the Mexicans ıtercalated thirteen days every fifty-two years. Still far- ıer: intercalation was proscribed in Egypt, to such a point ıat the kings swore, on their accession, never to permit ıem to be employed during their reign. Notwithstanding ıis difference, we find a very striking agreement in the ngth of the duration of the solar year. In reality, the ıtercalation of the Mexicans being thirteen days on each ycle of fifty-two years, comes to the same thing as that of ıe Julian Calendar, which is one day in four years; and ınsequently supposes the duration of the year to be three undred and sixty-five days, six hours.

"Now it is remarkable that the same solar year of three undred and sixty-five days, six hours, adopted by nations different, and perhaps still more remote in their state of vilization than in their geographical distance, relates to a real astronomical period, and belongs peculiarly to th Egyptians.

"As to the Mexicans, it would be superfluous to exam ine how they attained this knowledge. Such a probler would not soon be solved; but the fact of the intercalatio of thirteen days every cycle, that is, the use of a year o three hundred and sixty-five days and a quarter, is a proo that it was either borrowed from the Egyptians, or that the had a common origin."

Architectural Evidence.—In noticing these seve ral heads we can not do it more briefly than in th terms of the author, and in selecting the paragraphs to give a general outline of his proof, are at a grea loss which to pass over, and which to transfer. T give full force to his opinions, a complete transcrip would be necessary.

Fronting p. 55, we see a plan of the palace o Mitla, in Mexico, with ramparts and mounds.

The distribution of the apartments bears a striking an alogy to what has been remarked in the monuments of Up per Egypt, as drawn by Denon and the savans of the Insti tute of Cairo. Nay, the building itself is in the form o the Egyptian Tau.

In North America, the sepulchres of the ancient rac are the tumuli of the country. In Peru they are the same "The Indians, having laid a body, without burial, upon th ground, environed it with a rude arch of stones, or bricks and earth was thrown upon it, as a tumulus, which they cal *guaca*. In general, they are eight or ten toises high, an about twenty long, and the breadth rather less; but som are larger. They are in shape not precisely pyramidal, bu more like hillocks. The plains near Cayambe are covere with them; one of their principal temples having bee there, where the kings and caciques of Quito were buried."*

In the North American tumuli, various articles are foun buried with the occupant, such as idols, clay masks, mica

* Ulloa, Vol. I, p. 366. Gent.'s Mag., Vol. XXII, p. 210.

tone axes, silver and copper rings and rosaries. Precisely imilar articles are discovered in the sepulchres of Mexico nd Peru.

"In the tombs of Siberia, and the deserts which border t southward, are found thousands of cast idols of gold, silver, copper, tin and brass. Some of the tombs are of earth, nd raised as high as houses, and in such numbers upon the lain, that at a distance they appear like a ridge of hills." *

The most ancient pyramids of the Mexicans are those f Teotlihuacan, and are said to have been built by the Toltec race.

"The group of Teotlihuacan is eight leagues north-east f Mexico, in a plain called Micoatl, or the "Path of the Dead." There are two large ones dedicated to the sun, Tonitiuh), and to the moon, (Metzli), they are surrounded y several hundreds of small pyramids, which form streets, n exact lines from north to south and from east to west. One is fifty-five, the other forty-four metres in perpendicular eight. The basis of the first is two hundred and eight netres in length. It is, according to Mr. Oteyza's measurement, made in 1803, higher than the Myceninuns, the third f the great pyramids of Geiza, in Egypt; and the length f the base is nearly equal to that of the Cephren. The mall ones are nine or ten metres high, and are said to be urial places of the chiefs of the tribes. The two large nes had four principal stories, each sub-divided into steps. The nucleus is composed of clay, mixed with small stones, nd incased by a thick wall of porous amygdaloid. This onstruction recalls to mind that of one of the Egyptian yramids of Sakhara, which has six stories, and which, according to Pococke, is a mass of pebbles and yellow mortar, overed on the outside with rough stones." †

The pyramids of Dgizeh, in Egypt, it will be borne in mind, are also surrounded by smaller edifices in regular order, and closely correspond in arrangement to what has been ere described.

* Rankin's Conquest of Peru, p. 238.
† Rankin's Conquest of Peru, p. 356.

"The greatest, most ancient, and most celebrated of the pyramidal monuments of Anahuac is the teocalli of Cholula. At a distance it has the aspect of a natural hill covered with vegetation. It has four stories of equal height. It appears to have been constructed exactly in the direction of the four cardinal points. The base of this pyramid is twice as broad as that of the Cheops in Egypt, but its height is very little more than that of Mycerinus. On comparing the dimensions of the house of the Sun, in Peru, with those of the pyramid of Cholula, we see that the people who constructed these remarkable monuments intended to give them the same height, but with bases of length in proportion of one to two. The pyramid of Cholula is built of unburn bricks, alternating with layers of clay." *

This edifice, it would appear, closely corresponds with the great temple of Bel, or Belus, at Babylon, as described by Herodotus.

From this may we not learn the intention of the embankment around the large tumuli of North America: for instance at Circleville and Marietta? And do we not clearly see that this race continued the same manner of constructing their "high places" in Mexico and Peru, with the improvements incident to their permanent location there?

Another feature presents great analogy. Their buildings, particularly the sacred houses, were covered with hieroglyphics. Each race, Egyptian, Mexican, and Peruvian recorded the deeds of their gods upon the walls of their temples. Nay, science was also sculptured thereon, in both countries, in the form of zodiacs and planispheres, corresponding even in signs.

This section upon architecture, from which we have so freely taken, concludes with a lithograph of an image which Humboldt considers as an Azte princess. It was taken from the ruins of a teocalli at Tenochtitlan, destroyed by Cortez, and with th

* Essai Politique sur la Nouvelle Espagne.

exception of a string of beads across the forehead, is a tolerable copy of the Egyptian Isis.

Manners and Customs.—In the valley of the Scioto, in Ohio, and at several places in Kentucky, in the vicinity of ancient works, the "pyrula perversa" has been found in numbers exceeding one hundred, and generally at some depth in the ground. It is a shell, in size varying from six to fourteen inches in length, and when not injured is entire, and without artificial openings and marks. No such shell is known on the American coast, except a small specimen in the Gulf of Mexico; but they are said to exist as a marine production on the shores of Hindoostan, and are there used in the performance of religious ceremonies. Many other similarities are pointed out as having existed in Egypt, Hindoostan, China, and Peru.

A few lines may be well introduced here to connect Hindoostan and Egypt.

"The sepoys who joined the British army in Egypt under Lord Hutchinson, imagined that they found their own temples in the ruins of Dendera, and were greatly exasperated at the natives for their neglect of the ancient deities, whose images are still preserved. So strongly, indeed, were they impressed with this identity, that they proceeded to perform their devotions with all the ceremonies practised in their own land.

"But the most striking point of resemblance between the inhabitants of Egypt and India is the institution of castes—that singular arrangement which places an insuperable barrier between different orders of men in the same country, and renders their respective honors, toils, and degradation strictly hereditary and permanent."

The author ought by no means to omit to state that precisely the same division of caste prevailed among the ancient Mexicans and Peruvians.

Here terminates the comparison between the ancient inhabitants of southern Asia, and south and central America, for the purpose of sustaining their identity. The next step is to trace a connection *by emigration*, from Babylon to Egypt, Scythia, Siberia, North America, Azatlan (western States), Anahuac, Mexico, and Peru. Our object is less to criticise than to present an outline of this extraordinary work; extraordinary, not so much on account of originality, as judicious and extensive research; as presenting an accumulation of sensible matter relevant to this interesting question; as embodying all the learning extant, which has a rational bearing upon the origin, superstitions, and general character of a race whose deeds are so thoroughly obscured by the lapse of time.

On page sixty-eight, the author says, "we now enter the most difficult, yet the most interesting part of our subject, the endeavor to *trace* the origin and history of the aboriginal race of America. Cuth, Cush, or Chus, the grandson of Noah, and son of Ham, was the ancestor of the Cuthites, who built Babel. This took place under Nimrod, the fourth from Noah, and after the dispersion consequent upon the confusion of tongues, he founded the *Ancient Scythian* Empire; Scythian being the Greek style for Cuthite.

It appears the warlike subjects of Nimrod and their descendants gave their names to all countries conquered or occupied by them, and that the same name has, since their departure or emigration and from the time of Herodotus to this day, been used to designate a different race and country. The Tartars and northern Asiatics, or Mongolians, who occupied Dacia, and the Caspian, are also called Scythians. This is an important distinction. After the confusion of language, "the country about Babel was evacuated. A large body of the fugitives betook themselves to Egypt, and are commemorated under the name of the Shepherds."—Bryant's Ancient Mythology, Vol. III, p. 262.

"These Cuthites, then, obtained the mastery of Egypt, established a noble empire, under the title of "the Shepherd Kings," and constructed, as they did in Chaldea, large cities, pyramids, obelisks, and other massive buildings, the remains of which still furnish testimony to the magnificence and power of the race. "The Shepherds are said to have maintained themselves in this situation for five hundred and eleven years. At last the natives of Upper Egypt rose in opposition to them, and defeated them under the conduct of King Halisphragmuthosis. They afterwards beleaguered them in their stronghold, Avaris, which seems to have been a walled province, containing no less than ten thousand square arouræ. Here they maintained themselves for a long space; but at last, under Thummosis, the son of the former king, they were reduced to such straits as to be glad to leave the country."* "Wearied out by the length and straitness of the seige, they at last came to terms of composition, and agreed to leave the country, if they might do it unmolested. They were permitted to depart, and accordingly retired, to the amount of TWO HUNDRED AND FORTY THOUSAND PERSONS. Amosis, upon this, destroyed their fortifications and laid their city in ruins."

Early writers notice the journeyings of this banished race in a north easterly direction as far as Palestine. Here all historical traces are lost of them, and their name is buried in oblivion.

On page seventy-five will be found the following paragraph, relative to various emigrations from Egypt:

There were no less than three exodi from Egypt. The first was the one just named, viz: the expulsion of two hundred and forty thousand Cuthites by Halisphragmuthosis; this occurred about two hundred years before the entrance of

* Bryant. Vol. III, p. 237.

the Israelitish shepherds into Egypt. The second exodus was that of this once holy people, under the guidance of the Almighty, through his servant Moses, the account of which we have in profane history, substantiated in the minutest particulars by the sacred writings given us through the inspiration of the Holy Spirit, which protected and preserved the race. But the third is not so generally known. We propose to give a statement thereof and show the authority on which it rests. The author deems it necessary this should be kept in view, in order to prevent doubt as to the course taken by the first emigrants from Egypt.

And at page eighty it is said:

From what has here been related, then, it is thought that little or no doubt can arise as to the destination of the three expelled races, on their departure from Egypt: *The first*, in a north-easterly direction through Palestine; *the second*, under Jehovah's guidance, into the land of Canaan; and *the third* through Greece, westwardly through Europe, to their final destination in Great Britain. Here, then, we return to the subject matter of our investigation, viz: the progress of this first migratory race of Cuthite "shepherds," after they journeyed from Egypt to Palestine.

Pursuing the chain of reasoning with the closeness of a legal argument, the author says:

It will be recollected, perhaps, that in the argument exhibiting the anatomical analogy between the aboriginal race of America, and that of Southern Asia, a close affinity was marked between the characteristic traits of the North American Indian, and the Mongul or Tartar race, in their nomadic life, and their rejection of civilization. We find in North America, tumuli, ramparts, etc., which the Indians know nothing about; and from what has thus far been shown, these works prove to be the remains of some other, and a more civilized race. The Mongolian family are equally rude vith the Indian, and as little disposed to exert a talent for nechanical ingenuity. If, then, we find in Tartary and Si›eria monuments like the American, displaying industry .nd talent, unknown to and unpractised by those nations, ve must necessarily conclude they are the works of some .ncient and great people once occupying the land so enriched ›y the remnants of former greatness and power. That these xist, it is proposed to show:

"In the museum at St. Petersburgh, are preserved a aultitude of vessels, diadems, weapons, military trophies, rnaments of dress, coins, etc., which have been found in he Tartarian tombs, in Siberia, and on the Volga. They re of gold, silver, and copper.

"In the tombs of Siberia, and the deserts which border t southward, are found thousands of cast idols of gold, siler, copper, tin, and brass.

"Some of the tombs are of earth, and raised as high as iouses, and in such numbers upon the plain that at a disance they appear like a ridge of hills; some are partly of ough hewn stones or of free-stone, oblong and triangular; thers of them are built entirely of stone."

When, then, we find history pointing us to an exiled ace, slowly travelling in a northwardly direction, through iosts of foes, whose animosity, revenge for past tyranny, .nd spirit of self preservation, would constantly drive them orward and onward; and when we see this race possessed f the very genius, which, in no other in those days, prouced a similar degree of excellence, enabling them to raise ›yramids and cities, and ramparts for protection, preserving heir dead with scrupulous care, and interring with them .uch animals and relics as were supposed to be of use in a uture world; are we not irresistibly led to the conclusion hat this family arrived at this land, and for a season held lominion over it?

From the analogies comprised in the early portion of this vork, we also clearly see that some ancient race came from he southern parts of Asia, and, wandering southwardly hrough America, resumed their ancient customs, preerved in some degree their language, built ramparts, pyramids, and cities as of old, and established their primitive systems of mythology and astronomy. History, too, points out clearly the emigration from Babylon to Egypt, Egypt to Caucasus, and Caucasus to Siberia, of a learned, war-like, and great nation. We also know they were driven hence, but here we lose all traces of them, and their only vestiges are the walls and ramparts, tumuli and medals, yet discoverable in the latter country, where, since their time, a nomadic race, and one partaking in no degree of the excellence of that driven away, has held dominion. From the analogical evidence alluded to, there is some probability they went to America from Siberia, and founded the civilized empire there discovered. It is deserving of inquiry, whether this probability can be made a matter of certainty. In order to do this satisfactorily, it were well first distinctly to understand the position and distance of that very narrow passage of water dividing Asia from America, usually known as Behring's Straits.

The *practicability*, then, of a passage across these straits is made certain. *They are only fifty-two miles across, and that distance is divided by three islands.* To establish, then, the *probability* that emigration followed this route, the following considerations are worthy of attention:

And here a new piece of evidence unexpectedly presents itself, consisting of a *hieroglyphical map*, obtained in Mexico by Chevalier Botturini about 1780. The Chevalier was imprisoned, lost his papers and died in confinement, of a broken heart. Mr. Bullock, formerly of London, now of Cincinnati, afterwards visited Mexico as a traveller and antiquary, and was sufficiently fortunate to recover the "Aztec Map" of Botturini, which is engraved and prefixed to Mr. Delafield's work, with the remarks and notes of Botturini in fac simile.

If it is authentic,—and its correspondence with this kind of Mexican representation, the character of those who vouch for it, the manner in which it was obtained, leave, it seems, *no doubt* upon that point,—a most striking corroboration of the supposed journeyings of this people is then presented.

The native Mexicans stated it to be a chart delineating the entrance into America of the Aztec race, and a narrative of their slow and polemic journey southwardly into Anahuac.

It commences, as they alleged, with the departure of their ancestors from an island.

The drawing begins by exhibiting an enclosure, intended to designate the boundary of *a narrow passage of water, in the centre of which is an island*, and from this island they reached the main land in a boat, as is here portrayed. On the island are six hieroglyphics each, denoting the word "*calli*" or "house," surrounding the emblem of a tumulus erected for worship. Beneath are two figures, male and female, the latter being distinguished by the two small tresses resembling horns, as in the mythological painting of the age of famine. Attached to the female is an emblem used heraldically, and points her out as one of the "children of the sun,"—a title claimed equally by the Hindoos, ancient Egyptians, and Peruvian Incas.

Here it were well to notice how distinctly it is shown that the emigration into America of this civilized family, was from an *island in a narrow passage of water*. Search the continent on all its coasts, and no such place is to be found except at Behring's Straits, which have been already described. Is there not, then, additional proof in confirmation of the opinion, that this passage was that which facilitated the peopling of America from the nations of Asia.

For a full idea of this map, as describing the progress of the ancestors of the Mexicans, we can only refer to the work itself.

It were impossible at the present day to exhibit the positions of the various towns, which we find delineated on this map. No doubt the traveller through the north-western part of the United States passes them constantly. Here he meets the ruins of an ancient city, of which nought remains, save its ramparts and "high places," and there the lofty tumulus and range of walls point out to him the spot where sacrifices were once offered, or beacon fires were lighted. The names, however, and glory of those places have departed, and they are an enigma to the world. This map, no doubt, gives us the appellation of the most prominent cities, but to locate them with certainty were beyond the power of the present age.

"The author concludes the book by an abstract of the points made and the testimony offered in support of them, and at the finis thus takes leave of his readers.

Do these incidents form a well connected chain?

The evidence adduced is no *hypothesis*. It is based on the testimony of the most credible witnesses, whose names and works have been cited in their respective places. The author omits any argument on the premises, and deems it unnecessary. With the simple statement, then, of recorded incident, he submits the case to the candid and courteous consideration of the reader; and to him he tenders a respectfully and probably a final farewell.

We have already extended this notice so far, that comment will render it tiresome. It is not every position of the work that will bear examination; in fact it would be very strange, if, in the prosecution of an inquiry where all *ordinary* lights are put out and the darkness of at least twenty centuries has succeeded, one-half or one-fourth of our author's propositions were not open to doubt. As an instance, it is highly improbable that a people who knew the use of iron, as it seems the descendants of Cuth did, would ever lose it. The stone tumuli and walls of the North exhibit no evidences of having been wrought, and the universal belief is contrary to the supposition that this metal was known here. But we cannot too much admire the soundness of manner and the logical precision with which the subject is treated. It contrasts in this with many similar productions of antiquaries. In endeavoring to penetrate the obscurity which shrouds the transactions of the past, most men become bewildered, and indulge in speculations which require more credulity than reason to be received as truths.

ON THE NATURAL TERRACES AND RIDGES OF THE COUNTRY BORDERING LAKE ERIE.

[American Journal of Science, July, 1850.]

THROUGH the assistance of the engineers, engaged at various times in surveys for railroads and canals in Northern Ohio, I have been enabled to determine the elevation of our "Lake ridges" at numerous points between the Pennsylvania line and Sandusky Bay, a distance of 130 miles. I am more particularly indebted for these levels to J. H. Sergeant. Esq., who has run several lines west of Cleveland, and to Messrs. Harback and Smith, engineers for the Cleveland, Painesville and Ashtabula Railroad Company.

When these surveys do not cross the ridges and terraces, they have still been the basis upon which by short cross levels, taken with a pocket instrument, I have obtained the elevations; and the results, I think, cannot be wide of the truth.

There may be an extreme discrepancy of three feet among them, however, arising from changes in the surface of the lake, which is the common plane of reference.

My opinion has been for many years, that the "*ridges*" are not "ancient beaches" of the lake, although some of the *terraces* may be. It is indispensable to a beach that it should at its foot or water line be *perfectly horizontal.* The lake ridges are not so; and this fact, taken with the external form which they assume, clearly gives them the character of *sub-marine* deposits.

There are points on this coast where there are four ridges rising in succession from the lake, as in the township of Ridgeville, Lorain County. In other places there are *three*, as from Geneva to Ashtabula; from Euclid through Painesville to Geneva, two; and from Cleveland to Euclid, one. There are places where it is difficult to trace any; and in others, as at the city of Cleveland, where there are two or three branches or divisions of one ridge for short distances, all about the same level and liable to terminate suddenly. The ridges are sometimes upon the crest of a terrace, and sometimes lie, like a highway of water-washed sand, on the gently inclined surface of a plain that descends toward the lake. From a regular and beautiful elevated road-way, the ridge occasionlly breaks into sand knolls, as at Avon Centre, Lorain County; at Ohio City, near Cleveland, and at Painesville, Lake County.

Where nothing to the contrary is stated, the height given is that of the summit of the ridge, terrace or knoll. The first ridge, or that nearest the lake, is known in the county as the "North Ridge." The others have different names at different places; as the "Middle Ridge," "Chesnut Ridge," "Butternut Ridge," and "South Ridge."

Elevation of the North Ridge, beginning at the Eastern part of the Western Reserve.

	FEET.
Conneaut, Ashtabula Co., above Lake Erie,	120
One mile west, ...	145
Four miles east of Ashtabula village, .	132
(Base of same for several miles, 85 to 95.)	
County line between Lake and Ashtabula Co., northern slope of North Ridge .	107
Eight miles west in Lake County, .	125
Centreville, one mile north of village, .	105
Painesville, ...	120
Mentor—well defined for two miles level,	109
Willoughby, ...	85
Seven miles east of Cleveland, .	112 to 118
Three " " " " . . .	113 " 118
Two, " " " " at crossing of Cleveland and Pittsburgh Railroad, ...	128
Cleveland City, ...	96 to 108
Ohio City, ...	114
Rockport, Rocky River, ...	90
One mile west . . .	105, 107 and 126
Avon, Lorain County, east of Centre one mile, ...	84
" Lorain Co., Centre sand knolls,	105
Russelton, Lyme, Huron Co., . .	120

This table embraces a distance of one hundred and twenty (120) miles, where it appears the lowest summit is 85 feet, and the highest 145, showing a difference in longitudinal direction of 60 feet. I have not visited all the positions here given, but the greater part of them, and for the rest am informed by the engineers that there is no higher ground between the ridge and the Lake. In all cases there is a smooth uninterrupted plain, on the lake side, over which the water of the lake is every where visible when the forest timber, which is heavy, is cleared away. It is variously composed of blue marly clay, of coarse drift, called "blue" and "yellow hardpan," and of coarse sandy and gravelly drift; but the soil is for the most part clayey, and wet between and below the ridges. The streams, little and great, cut deep and steep gullies through the superficial deposits, and also into the rocks below. From the cliff limestone at Sandusky, eastward and to the State line, the superficial matter rests on slates, sandstones and shales, corresponding to the Hamilton, Chemung and Portage groups, of the New York reports.

Elevation of the second Ridge, called the "South" and "Middle" Ridge.

	FEET.
Near Kingsville, Ashtabula County (south ridge),	152
Centreville, ... "	122
Two miles east of Cleveland, . "	135
Two miles south-west of Ohio City (middle ridge),	149
Dover Centre, 12 miles West of Cleveland (middle ridge),	163
Rockport, seven miles west of Cleveland (middle ridge),	130
Ridgeville, Lorain County (middle ridge),	168

This ridge is more broken and less continuous than the first, or "north ridge," and is in general heavier. In Rockport, Dover, and Ridgeville, on the northern or lake slope, it is from 16 to 20 feet above its base or foot, and on the rear 5 to 10 feet. Behind it, as with all the ridges, is flat, swampy land, and small rivulets that drain the low ground, running parallel with the swell, to some creek, or occasionally breaking through towards the lake. These lands are very rich, and with a moderate expense are drained by ditches cut through the ridge. The slope of the flat lands between and before the ridges is sufficient to carry off all the water in ditches that have a free current. Most of this land is coming under the plough in this manner, although it is equally well calculated for grass. No country can possess more rural beauty than that along these "ridge roads." The land in a longitudinal view, is apparently level as far as the eye can reach; and the buildings congregated along the line of the road appear to be arrayed in curved lines, gently waving to the right and left as you proceed. Looking from one of the interior ridges, which are generally perceptibly higher than the next one towards the lake, if the timber is not standing, another and rudely parallel row of farm houses, barns, orchards, &c., is seen at the distance of one, two, or three miles; the intermediate space perfectly smooth and cultivated, and beyond lies the blue water, and the horizon.

The composition of one ridge does not materially differ from another. It is formed of coarse, water-washed, yellowish sand, or of fine gravel, principally the comminuted portions of the adjacent rocks. The rocky fragments are not generally worn perfectly round, or oblong, as beach shingle is, but are more flat, with worn edges. There are mingled with the sandstones and shales, that compose this gravel, scattered pieces of quartz, flint, also granite, and trappean rocks, limestone and ironstone.

The basis of the ridge corresponding with the impervious clayey soil between, gives rise to a great many springs on the lake or lower side; and this water frequently deposits bog iron ore that has been used extensively in furnaces along the lake shore.

From near Dover centre, west of Elyria, and even to Vermillion River, the second or "middle ridge" rests on a coarse-grained sandstone or "grindstone grit," which farther east in Cuyahoga County rises above the level of the lake ridges. Between the Black and Vermillion rivers I have not succeeded in procuring the elevations. Here they are well developed, and show more branches or collateral lines, extending from one ridge to another, than is observed farther east.

By digging shallow wells, the inhabitants find water in abundance, and generally good. In these wells, from 12 to 18 feet deep, there are thrown out as a common occurrence, sticks, timber and leaves, in a decaying state. I have in my cabinet some pieces of this wood, furnished by Dr. Moore of Dover, who took it from a bed of carbonaceous matter in a well of his, twelve feet below the surface. The well is situated on the middle ridge, 163 feet above the lake. Pieces of timber six inches through have been found, represented as being water-worn like drift-wood. Those in my possession are solid, with a very fine grain resembling the willow.

Dr. Moore, an intelligent physician of my acquaintance, says he has seen shells thrown from the the bottom of wells, resembling "perriwinkles," a comcommon name for Lymnea. Similar shells in fragments are said to have been thrown from a pit two miles west of Cleveland, on the north ridge. In the "blue marly clay," beneath this ridge, I have found a Helicina, and a Planorbis, shells characteristic of the loess of the Rhine, and of St. Louis, and the Wabash in Indiana. The palæontological evidence is therefore, as far as it goes, in favor of the idea of very recent and fresh water deposits.

It will be seen from the second table of heights, that the greatest difference is there forty-six feet; and that the summit of the ridge rises, from Rockport to Dover, in seven miles, thirty-three feet, but from Dover to Ridgeville, six miles, it is nearly level.

Through these two distances, making thirteen miles, the height of the ridge above its base is about the same, from sixteen to twenty feet; and consequently the base has an equal rise in a longitudinal direction. Two miles west of Ridgeville Centre, the top of the middle ridge has descended from 168 to 149 feet. The foot of the north or first ridge, and of the terrace on which it is frequently situated, approaches nearer to a horizontal line than the ridge itself, but still differs from a perfect level, It is at

	FEET.
Conneaut Creek,	75
Four miles east of Ashtabula,	85
Several miles west of "	95
Painesville,	85
East of Willoughby, several miles,	60 to 65
Three miles west of Willoughby,	60
Euclid Creek, 12 miles east of Cleveland,	75 to 85
Seven miles east of "	105
Two " " "	102
Ohio City,	75
Rockport,	70
Avon,	70

It is not easy to determine with precision where the base or foot of a ridge graduates into the plain; and consequently there is not that accuracy in the elevations for the base just given, that we attain when measuring the summit or crest. But they are a close approximation, and although remarkably uniform, are by no means equal as they should be if the base of the ridge represented an ancient coast line; the greatest difference being forty-five feet, or about the same as the variation along the top of the second ridge.

There are but few measurements in my reach of the third and fourth ridges. In Huron County, south of *Russelton*, there are two low swells of land parallel with the shore, apparently about on a level with each other, and not much above the main ridge at Russelton, which is reported at 120 feet.

The third ridge, in Ridgeville, Cuyahoga County, is one mile southerly from the second or "middle ridge," and is not very prominent, rising six to ten feet above the low ground. At this place it is 186 feet above the lake, or eighteen feet above the middle ridge, and eighty-one above the highest part of the north ridge in Avon, five miles north.

The fourth or last and highest well defined Ridge.

	FEET.
2½ miles southwest of Ohio City, . .	173
1¼ miles south-east of Ridgeville Centre, .	203
West bank of west fork of Black River, Elyria, ...	195

Distance embraced, twenty-five miles.

The materials of the most southerly or interior ridge are in general coarser than in others, showing a more violent or less lasting aqueous action. This is observed everywhere at the west. The more *elevated the drift*, the more does it exhibit the effects of *strong currents* in the transportation of large pieces of rock, in the shape of coarse gravel. The lower portions, especially those that lie near the surface of the great lakes, not only on Lake Erie, but on Michigan and Superior, are fine, argillaceous or marly, laminated, and with few pebbles.

The terraces have not been as much noticed as the ridges, and consequently their height is not as well known. From Rockport to Avon, the north ridge is upon the edge of a terrace, the foot from seventy feet above the lake, down to sixty feet; its crest from one hundred and five down to eighty-five. Directly opposite this, about five miles more inland, a considerable portion of the fourth or south ridge, (known as the "Butternut,") is also on a terrace of about twenty feet, on its northern face; in fact all the ridges partake of the nature of terraces, in places; the northern slope being generally the longest. But the geological composition of the terrace on which the ridge rests is different, and either a rock or a drift of more compact and resisting kind.

Between Newburg and Euclid, nine miles, the northern face of the terrace is very bold, its base from 120 to 150 feet, and its crest 200 to 225 feet. It is here composed of fine-grained sandstone (Waverly), and blue and red shales. East of Euclid, the terrace sometimes divides into two, the lower one supporting the north ridge. It is the same for several miles east of Willoughby, the crest of the first or lowest terrace being about one hundred feet and its base seventy to eighty feet, and formed of blue hardpan, resting on shales. In Erie County, on the line of the Mansfield railroad, it is composed of cliff limestone supporting black slate; its base about 130, and its summit 180 feet.

It will be interesting now to compare the elevations above given, for ridges in Western Ohio, with those of the great lakes in other States. In Michigan, at the east line of Washington County east of Ann

Arbor, is a well-defined ridge running nearly north and south, whose summit is 140 feet above Detroit River, at Detroit. Around Monroe, in Wayne County, Michigan, are some irregular sand ridges, not more than thirty feet. They are also visible on the north shore of Lake Erie, in the flat country between Erie and Huron; their elevation is not known, but they are apparently as high as 200 feet.

Mr. Roy, a Canadian engineer, has made a section across the ridges back of Toronto to Lake Simcoe, as reported by Mr. Lyell, and has given their respective elevations as follows:

	FEET.
No. 1.—one mile north of Toronto, 20 to 30 feet high—base above Lake Ontario, ...	108
No. 2.—2½ miles from Toronto, 50 to 70 feet high—base above Lake Ontario,	208
No. 3.—5 miles from lake—10 feet high—summit, ...	288

Five other ridges *or terraces* are given by Mr. Hall, in the geology of the 4th district of New York; also on the authority of Mr. Roy, referring apparently to the elevation of their base.

	FEET.
No. 4.—above Lake Ontario, ...	308
" 5. " " "	344
" 6. " " "	420
" 7. " " "	680
" 8. " " "	762

Mr. Lyell observed *eleven* ridges between the lake and the summit for Lake Simcoe, the elevation of the eleventh, or last and highest, corresponding with No. 7, of the New York Report. The elevation of Lake Erie above Ontario is generally stated at 332 feet, so that the three first ridges or terraces, in rear of Toronto, are below the surface of Lake Erie.

Mr. Barrett, a New York engineer, furnished Mr. Hall with the height of some points on the Lockport ridge, south of Lake Ontario and opposite Toronto. They are as follows:

	FEET.
At Lockport, ...	158
Middleport, 10 miles east, . . .	185
Albion, Orleans County, . . .	188
Brockport, Monroe County, . . .	188

None of these correspond in height with those on the north shore, as they should do if they were the result of littoral action at a beach, for the surface of the water would be level. If we suppose them to have been formed in that manner, when the water stood at the base of a ridge, the rivers must have discharged at the same level. Here should, on that hypothesis, be found deltas, and evidence of bays or lagoons. The waters having settled away, at the present period the streams discharge at a lower level, their channels being worn deeper and larger, cutting through the ridges and terraces that lie between the present and the ancient level. If the ancient mouth was at a point different from that where the present channel cuts a ridge, it should be visible in the present form of the ridge. If it was at the same point, there should be marks of such action as always accompanies the meeting of running currents with dead water. But our streams appear to cut the ridges as though they were barriers preëxisting, and broken through by the current.

Terraces, composed of the rocks or other general deposits of a country, appear to be much stronger proof of ancient shores than limited sand ridges.

When we rise above 240 feet from Lake Erie, th well defined terraces disappear; and from that line t 600 to 650 feet, the general elevation of the tabl land in north-eastern Ohio, the surface presents a con fused arrangement of heavy drift, covering the rock at various depths, in long massive knolls, withou ranges or parallelism. Towards the west, the summit of the lake streams are lower, and the present surfac of north-western Ohio and northern Indiana, of Illi nois, Michigan, New York and Canada West, wit much of Wisconsin and Iowa, would be submerged b a sea rising 250 feet above Lake Erie, or 815 abov the ocean.

The Wabash and Maumee summit, at Fort Wayne Indiana, is 246 feet above Lake Erie. The summi between the waters of

	FEET ABOVE LAKE ERI
Saginaw Bay and of Lake Michigan, . .	108
Summit between Pishtaka and Rock River in Illinois, ...	218
Lake Winnebago, ...	167
Summit between Lake Ontario and Lake Simcoe, Canada, ...	197
Mouth of St. Peter's River, Fort Snelling, Minesota, ...	179
Missouri River at Fort Leavenworth, west line of State of Missouri, . . .	181

If, therefore, the relative level of the land was th same as now, when the diluvial sea existed at hig levels, its extent must have been very great, at th supposed stage of 250 feet above Lake Erie. At thi or any other supposable stage, if it remained station ary long enough to form cliffs and banks at one place it would produce the same effects, in kind if not i degree, at another; and we should be able to trac ɘaches or shores over all the vast west, for such cliffs ·e due to the action of winds and waves, always in)eration on bodies of water. But under-surface cur›nts are not universal, or in such general operation ı to form everywhere bars and longitudinal banks · spits. If the Atlantic Ocean should suddenly set-ɘ one hundred feet, or any other distance, would ıere not remain a distinct shore, well defined and aceable its entire length? At the mouth of some vers and bays or inlets would be seen limited sand dges, their bases upon an exact level. On the an-ent bed of the sea opposite sandy coasts, like North ɜrolina and New Jersey, would appear long and nar-w and rudely parallel ridges, of such materials as e easily moved by currents, that would *not* be level ngitudinally.

The evidence of the existence of ancient currents ting upon the drift, regularly and irregularly, is ›undant. They have acted at all elevations, as well the highest lands in Ohio, at 1350 feet above the le, as at the sources of the Mississippi, 1680 feet. ıe evidence is that they were powerful, and in gene-l erratic or irregular and fitful. Such currents ›uld not leave ridges, but rounded elevations. For is discussion it is immaterial whether the relative ange in the level of land and water was due to sub-lence or upheaval. The change has taken place, d before that period there were at great depths cur-nts of water, both gentle and strong, giving form the present exposed surface of the earth.

It is to this wide-spread power that we must resort, explain most of the diluvial phenomena which are served. What can be reasonably assigned to the aring action of waves along a coast line is limited d not pervading; not an universal, but a local, tar-, and inefficient geological agent.

THE AGE OF THE MATERIAL UNIVERSE.

[Western Quarterly Review, 1849, Vol. I, No. 3.]

Until within thirty years the general belief c mankind was, that the earth had been in existenc only about six thousand years. The researches o geologists, and the astonishing discoveries of astror omers have, within that time, entirely changed tł opinions of those conversant with natural philosoph and the natural sciences, on this point.

Men who have become proficients in the physic sciences now agree, the world over, that our eart and the solar system, and the yet unfathomed "stell world," have existed *hundreds* of *thousands* of year

The great past duration of the universe is an ide that new developments go continually to increas and not to diminish, so that we may safely estima its age at millions of years, instead of hundreds thousands.

I shall exhibit some of the facts upon which th belief is founded.

I may not succeed in causing so difficult a subje to be comprehended at once, but may give some a sistance in your future contemplations upon it.

The periods of time with which it is necessary deal are so vast, compared with our own earthly exis ence or even compared with the historic periods, th it is only by the use of our imagination that we can e brace the idea. Just as in Astronomy we are well convinced of the immense distance of the stars, but do not realize by comparison the length of a line extending from the earth to them, or even from the sun to the earth, 95,000,000 miles.

The distances which we travel on the earth, and over which ships sail on the ocean, being at most only 24,000 miles—or the circumference of the earth—become *as nothing* when applied to the heavenly bodies. The evidence of our senses in reference to space must be left behind, and the conclusions of the mind must be received as true.

I doubt whether any individual has a rational conception of the length of a line reaching to our moon, the nearest heavenly body, and distant only about ten times the circumference of the earth.

In contemplating the existence of the universe, we must accustom ourselves to grasp long periods of time, as in astronomy we do great extension of lines and spaces.

There is another difficulty in the discussion of the longevity of the earth, which arises from our prior convictions and opinions, and particularly from a fear on the part of some, and a belief on the part of others, that the doctrine here taught is contrary to the Holy Scriptures.

This branch of the subject is purely *theological*, and one the discussion of which belongs to theologians, and not to philosophers. But it is so interesting, and the apprehensions of Christians it seems to me so unfounded, that I am tempted to digress a little for the purpose of noticing its connection with revelation.

And in the outset I should observe—as I shall show before I close—that the science of astronomy presents far greater difficulties to be reconciled than that of geology, and the arguments in favor of the extreme past duration of our planet, drawn from ou knowledge of the stars, are the most conclusive of all

And I should add, moreover, that among the ge ologists and astronomers of my acquaintance, an those with whose personal history I am acquainted there are proportionally more devout and pious me: than there are among civilized mankind at large, o the secular professions.

The arguments which have been drawn from ge ology, unfavorable to the truths of Christianity, hav originated not with learned men. I know of no ex ception among the living geologists, of a man wh does not believe the earth to be more than an hundre thousand years old; yet at least one half of them ar exemplary, professing Christians, who assert that thei confidence in the scripture revelations is unimpairec

There was issued in England a few years since, work which attracted much attention, entitled "*Ve tiges of Creation.*" This book made pretensions t being an exponent of modern geology. But geolc gists of reputation in America and Europe regard th author's ideas as in the main contradictory, and nc in harmony with science—the vagaries of a disease or a half-instructed mind.

It is an exemplification of the words of the ol English poet and philosopher, that a "*little* learnin is a dangerous thing."

I know of no standard work on geology that at tempts to throw discredit on revelation, althoug nearly all of them date the creation of the world fa back of the historic period of 6,000 years.

The first and the most popular difficulty arise from *confounding* the era of the appearance of ma upon the earth with the creation of the world. Ther is no geological evidence, so far as I know, showin that man has been more than 6,000 years in existence

3ut the fact of a long prior existence of the earth tself is a *separate idea*, and should not be mixed with he *peopling* of the world by human beings.

There has been a great deal written upon the uestion of reconciling the Mosaic account of the cre- tion with the present state of science. The qualifi- ations necessary to discuss this question are possessed y few men living, if they are by any. It requires a nind free from every external and improper bias. 'he man, or the men who enter upon the discussion, nust be linguists of the first class, to scan the Mosaic ccount thoroughly in all the languages through which ; has passed, from the Hebrew to the English.

They should be masters of modern science, and ere it is doubtful whether the progress of learning is et such as to render it safe to make a final conclu- ion of the matter.

The translators of the Septuagint had only the stronomical light of the era of Ptolemy. The ranslators of the English Bible were a little better nformed, but even then the Copernican system was ot universally received. Great strides in scientific iscovery have been made since the days of James he Second.

But I have not forgotten my resolution to express ome of the modern views entertained upon the sup- osed discrepancy. It will be of course principally a epetition of what may be seen by you all in publica- ions upon geology, theology, and astronomy.

The first chapter of Genesis does not appear to ave been intended as a piece of *natural history*, and hould not be regarded as such. There is no doubt ut he who made all things knows all about them, and ould give us, if he chose, at once the laws of matter nd the most perfect history of the *manner* in which hey were originated. In our researches we should first divide the *act* of *creation* from the acts of *change* to which matter has been subject *since its creation.* The absolute *creation* of *matter* is an idea beyond our comprehension.

The prevailing theory is that the first or primeval state of the planets was that of a vitrified mass, which would naturally become round by motion on its axis. But how that mass was brought into being is not as yet a subject of human speculation, it is referred directly to the incomprehensible power of God. The observations of man have disclosed abundant evidence of *change in the condition* of our planet; and the *laws*, the *number*, and the *nature* of *these changes, constitute the science of geology.*

I conceive it is only the first verse of the first chapter of Genesis that may be truly said to refer to the *creation*, technically speaking:

"In the beginning God created the heavens and the earth."

The next verse describes the condition of the earth *after* creation.

"And the earth was without form and void, and darkness was upon the face of the deep," &c.

The succeeding verses are chiefly taken up by a statement of the *changes* introduced in the earth, which had already been made.

The dry land appears—oceans of water result from the emergence of the land—plants spring up—the waters bring forth things that have life. Here is past creation, and past modification of what was created. The animals, plants, and fishes were *originated;* the mountains and seas were the consequence of a rise of the solid materials of the globe, that is, its rocky portions, thus driving its waters into the great primeval valleys forming oceans and lesser bodies of water.

But I do not think there is an *attempt* in the first hapter of Genesis—or anywhere in the Bible—to ;ive a full, clear, and detailed description of these stu-)endous events.

The revelation is a moral and not a scientific one; vhat there is in it of science, history, or political aw, is *incidental* to the morality and religion there :ommunicated, and goes no farther. What is purely ecular is merely illustrative, and to be understood nust have been made in terms in use, and in accord- ,nce with the state of intelligence of the people o whom it was addressed; if not, it was to them no evelation—it would be merely nothing. The He-)rews had seen by night a revolving heaven filled with tars, and had so often watched these mysterious ob- ects as they presented themselves anew each evening hat they had discovered some which had a relative notion among the others.

Here their knowledge of the planets and the plan- :tary system ended. The people of Israel could ıeither read nor write. The priests were the de-)ositories of knowledge, and they were not much n advance of the people. What did Moses, their)rophet and legislator, know of astronomy or of ;eology? Was he inspired in matters of science :bove the learning of his day? There is no evidence hat he was. Had that been the case, it must have)een with reference to the moral revelations he was to nake to his people, and in them we should have found :xpressions relating to science. Was such a revela- :ion necessary to sustain and support the moral and ·eligious law? If it was, how much of it? If it was ·egarded by the Deity as of consequence to his pur-)oses, that mankind should be provided with an easy vay to secular knowledge, would he not have intro- luced such subjects in his revelation?

If it was necessary for his prophets to possess this knowledge, he would have imparted it to them, and they, recording his communications, must inevitably have made use of expressions that would show us that they had this knowledge.

The result would have been that a moral subject would at one time require or connect itself with mathematics, at another with natural philosophy, then with chemistry, and so on through the range of the sciences and the arts, not merely as known to us, but as shall be known through the progress of time, and as fully as known to the Divine mind.

Mankind would not have been in ignorance, till the days of Copernicus, that the sun occupies the centre of our system, and that the earth does not stand still in space.

It would not have remained for Kepler to show that the planets describe equal spaces of their orbits, in equal times nor to teach that the other great law of astronomy, that the cubes of their times of revolution are proportional to the squares of their mean distances; Archimides could have developed no new ideas on mechanical forces; Sir Humphrey Davy no discoveries in chemistry; Franklin, in electricity; or Morse, in telegraphs. We should have known of the existence of the planet Uranius before the days of Herschel; of Neptune, before Leverrier and Adams; we should probably know whether there is beyond the orbit of Neptune more than 4,000,000,000 of miles from our sun, another member, or other members of the solar system.

But these developments, not forming any part of the written revelation, I conclude that a knowledge of the sciences or of natural history was purposely kept from the books of revelation.

A query might be started, whether there are not

in fact some such matters contained in the sacred records? and whether, if there are, they must not agree with true science? So far as I am acquainted, I know of no subject *purely scientific*—no phrases that appear to be intended merely to give such information upon the pages of the Bible. If knowledge of this kind is anywhere advanced, it should, when fully understood, certainly correspond with our ideas on the same subject, where we understand it correctly and fully.

The divine reason and human reason are the same mental perception; and when humanity can reach the truth, its ideas correspond to those of the Deity on the same subject.

Thus the mathematical principle, that the sum of the squares of a right-angled triangle are equal to the square of the hypothenuse, is an idea comprehensible alike to the divine and the human intellect. So with the property of gravitation; with Kepler's astronomical laws; chemical affinity; geological principles, and all other knowledge. Our powers are limited to the *discovery* of God's laws; they are not equal to the origin of laws in nature or science. When the two grades of intellect unite, in the understanding of a subject, human reason, as far as it sees correctly, must be the same in *quality*, though infinitely less in quantity, as divine reason. The difference is, that we are continually subject to error; he is not: we are always learning; he never learns.

It is on such grounds that I regard the Bible as forming no part of our "natural history," or of our science; and that the first chapter of Genesis should not be regarded as a description or an attempt at a description of the mode or manner of the creation.

For such a purpose it is too meagre, too confused, and too blind; it would have required, not merely one short chapter, but many large volumes, to disclose as much respecting the act of forming and peopling the solar system, as we already know.

The most that can be drawn from that chapter, as bearing upon the *modus operandi*, is a succinct, unqualified statement—brief as it is sublime—that in some remote, undiscovered period God made the universe.

After the lapse of an indefinite period, a succession of *changes* and *modifications* took place, accompanied by additional acts of creation which may still be going on. The intervals between these periods are unequal and indefinite, or as yet not known.

The most learned theological commentators teach us that the Hebrew word, translated day, should be regarded as meaning epoch, era, period, generation, or time, and not a natural day of twenty-four hours; nor does it mean periods of equal duration. This construction is evidently necessary to accord with observed facts. The lower animals lived, died; and whole *species* disappeared and ceased to exist before others were created or brought into being. They could not, therefore, have been made simultaneously in the same day, unless they were also cut off instantly, and new forms of life put in their place through numberless repetitions. This would be a mere act of creative sport, without design or beneficial purpose, an act of which no one supposes the Creator capable.

Dr. John Pye Smith says, the first verse refers to an "indefinite epoch in past eternity;" also that the term translated "*created*" may be used for the words "adjusted" and "finished."

Dr. Jenks says, the Hebrew word "erets," or earth, "implies everything relating to the *terraqueo, ærial* globe; that is all that belongs to the solid and fluid parts of our world, and its surrounding atmosphere."

The expression "And the evening and the morning were the first day" has been rendered "and there was evening and there was morning the first day," and also, "and there was dusk and there was dawn the first day."

I said the description was too *confused* to convey a clear idea of the process, taking the order of narration for the *order* of events.

There is day and night spoken of before the firmament, and before the sun, moon, and stars. In the original the firmament is not created, but made to appear. On the sixth day man and woman are said to have been created, male and female. But in the second chapter, the creation of man, the formation of the garden of Eden, and afterwards the creation of woman, appear to have taken place, if it was in the order as mentioned after the seventh, or day of rest. This want of order and perspicuity is common to ancient writings and to the Scriptures, where rhetoric and the grammatical structure of sentences give way to the native vigor of untutored and imaginative writers.

Plants are said to have been created before animals and the inhabitants of the seas. As to land plants, the researches of geology, in its present state, give a contrary order, and show us that the mollusca existed long before these vegetables, but future investigations may show facts to change the received opinions of geologists. Not long since marine plants were supposed to be more recent than the mollusca, but the New York geologists found and recently published sketches of marine vegetables in the Potsdam sandstone, the oldest rock of the New York system.

I present these apparent discrepancies to show that there are difficulties in reconciling the various parts of the narration *to itself*, considered as a piece of natural history. And to enforce my belief that it was *not so intended.* Any enlightened human mind which should understand such a memoir, would study to make it clear, orderly and consistent, much more the divine mind.

Geology, however, discloses a good general resemblance in the order of *the changes* and creations to those set forth in Genesis.

Almost the lowest fossil remains, which must of necessity be almost the oldest in the order of chronology, have eyes fitted to receive light. This is the case with trilobites found low down in the "silurian system," though the lenses composing the eye are constructed of great power, adapted to a very dim light. These creatures crawled on the bottom of ancient seas, and of course were made *after* the creation of light, or their eyes would be useless. Light must also have preceded the growth of plants.

Fossil remains prove that animals inhabiting the waters were in existence before those that walk on land.

The fossil elephant, mastodon, buffalo, megatherium, &c., are more recent than the fossil saurians, or water lizards, which were of great size.

Geology establishes the position that the first condition of this planet is aptly described by the statement that it was "without form and void."

The crystalline rocks, sometimes called the "primitive rocks," such as granite, sienite, and gneiss, are regarded as the cooled crust or exterior of the earth, its central portions still remaining in a liquid or volcanic state. The surface of the earth before the "sedimentary" or stratified rocks were formed would present the appearance of confused heaps of fractured angular primitive rocks, now represented by the peaks of the Andes, the Rocky Mountains, the Granite Hills of

New England, the Alps, and the Himalayas. In the "Plutonic" or primitive rocks, there are no fossils; no evidences that during the long interval between the time when our planet took its place in the solar system, and the period when the sedimentary rocks were deposited, any living, moving, or breathing thing existed upon the earth.

A change then took place, and a new class of rocks began to be formed by the agency of water. Those rocks which we call the "sedimentary" rest upon the Plutonic, and contain everywhere the preserved bodies of living organized creatures. Those that are first in point of time, or lowest in the system, are a small shell-fish called the "Lingula." As we ascend in the geological structure, the fossil remains become more numerous, and new forms of animal life constantly appear, such as fishes, reptiles, and birds.

Again, geology teaches that the animals adapted to the use of man, and at the same time to live on dry land, did not come into being till the geological changes were nearly complete. It shows that they were made near the last act of creation, and either contemporary or closely connected with the production of man. The remains of the ox, horse, elk, deer, &c., are found in the last deposit of all the "drift," or as it is sometimes called the "diluvium." This is the mixture of sand, gravel, bowlders, clay, and earth, which covers the surface of the indurated rocks. But even here no remains of man are found of a fossil character, showing that he appeared after all the organic changes had been made, and the surface of the earth had taken, in general, its present form.

Thus it may be shown that geology finds in the progress of the creative acts, an order corresponding in a general way, with the Mosaic Books, and that the descrepancies are not greater than would be anticipated when we consider that it was no part of the sacred historians duty to teach cosmogony.

I now dismiss the subject of discrepances, to return to the main question—that of the proofs of great age in our earth. An entire volume would not be too much space to devote to it. I begin with our own region.

If you look about from the upland of this village,* you observe a plain, the general level of which extends from the Summit Lake northerly to old Portage and Cuyahoga Falls. On each side of this space are rocky highlands, in the township of Portage on the west, and Tallmadge to the east. The Cuyahoga river comes in from the north-east, the Little Cuyahoga on the south-east, and on the south the branch which runs from the Summit Lake through this village. These streams have worn channels or gulfs into the plain, from fifty to one hundred and fifty feet deep, and the inquiry arises at once, how long would it take their waters to do this work ?

The present course of our streams was marked out as the ancient waters receded, and the dry land appeared. Our streams are merely the conduits for water falling from the clouds upon the surface of the earth.

As the land emerged from the ocean, there would be at first short rivulets like those of an island, and as the waters fell the length and number of streams would be increased. These would unite, always taking the lowest ground, forming great trunks like the Ohio, and finally like the Mississippi. The mouths of course changing position, and advancing as the level of the sea into which they discharged, became lower and lower. This is on the supposition of a gradual

* Akron, Ohio, where this discourse was delivered as a Lecture before the Lyceum, March 7, 1849.

subsidence of the waters, or a gradual rise of the land, both of which may have taken place. In some parts of the world, these movements were convulsive and not gradual.

Imagine the level of the water to be at the summit of the bluffs at the old Portage; and, as it settled away, the Cuyahoga beginning its work of excavation upon the drift which occupied its immediate valley.

The deposits of sand, clay, and gravel, would be rapidly removed, as many of you have witnessed, in the artificial grades effected by means of water, at the sand bluffs of the Little Cuyahoga, north of this town.

But beneath this drift, there is a rock to which the streams soon come; and after that their progress in excavation must be slow, depending upon the nature of the rock for hardness. The surface rock here is called "conglomerate," or the "pebbly sandstone," being a coarse-grained sand rock, with white silicious pebbles imbedded. It is on the average about one hundred feet thick, and may be seen in the street just north of the Court House, at the canal quarry, at the point of rocks, the old forge, Middlebury, and in both banks of the Cuyahoga, from the Valley Forge to Cuyahoga Falls.

The rock below it is composed of alternate layers of soft shale, and close-grained sandstone, and iron ore.

The fall or descent of the two Cuyahogas is about one hundred and sixty feet in two miles along the rapids: more than half of which is in chutes, and tumbles over the conglomerate rock. The fine-grained sandstone, and shale beneath, in which the channel is formed below the edge of the conglomerate, is easily worn away by the force of the current and by frost.

On the contrary, the conglomerate is a very durable rock, not easily disintegrated by weather or by water. The characteristic features are those of cliffs, standing out sharp cut to the day, while softer rocks form rounded and earth-covered bluffs. Wherever this rock is found, there are water-falls, and a rude, wild scenery. When the waters of the great Cuyahoga begin to fall over the edges of this rock, at the lower end of the gorge at Valley Forge, the shale beneath would be torn up rapidly by the force of the descent: but the conglomerate, acting as a protection, prevented the channel from wearing up stream as fast as it would have done if the rock had been all shale. The *undermining* process would go on below, and at the same time the wearing process above. There is now presented a chasm two miles in length, at the lower end about 200 feet deep, at the upper 25 or 30; its width but little greater than its depth.

How many years must have elapsed during the formation of this gulf? Along the upper half the water has not yet cut through the hard conglomerate rock, which there forms the bed as well as the walls of the gulf. Since the occupation of white men, no perceptible gain has been observed upon the rocky channel of the river by the action of the elements. It is a rock, which, when quarried and put in buildings, bridges, and walls, is regarded as proof against weather and climate. Its constituents are silicious crystals and pebbles, cemented by a little iron. And yet the never-ceasing flow of the Cuyahoga is eating a pathway through its mass at a slow but certain rate.

If we could, by long observation, determine how much was gained in a period of years, as they have at the falls of Niagara, we could give the estimate a mathematical form and certainty.

At the Niagara river, the chasm is seven miles in length, and the most eminent geologists have made computations upon the rate of its advance. They iave a map of the figure of the Fall, made by Hen-iepin, about 1680, and have studied the character of he rocks which lie beneath.

There, as here, there is a substratum of shale, asily torn away; the lime rock at the crest of the 'alls being of better resisting power. They have lso discovered an ancient channel which leaves the resent one at the whirlpool, and reaches Lake Onta-io some miles west of the present mouth of the Niag-ra river. This channel is now filled with drift; but hat the river, or at least a part of it, once discharged here, is well established by Sir Charles Lyell. The eologist of New York, Mr. James Hall, has made lose surveys of the Fall as it now is, and in addition o the map has fixed copper pins in the rocks, by which neans future examiners may judge of the recession f the Falls.

After all these investigations, I believe no geolo-ist fixes the period that must have been occupied by he Niagara river in excavating its present channel elow the Falls, at less than 43,000 years.

Considering the resisting nature of the conglome-ate as compared with the Niagara limestone, which s slightly soluble in water, the less mechanical power f the stream, and the apparent stability of the rock ince the settlement of this country by whites, I am nclined to regard the time necessary to wear out the ulf of the Falls of Cuyahoga as great as that for Niagara, or say 50,000 years.

This is one item in our computation, and applies nly to the lapse of time, since the deposition of the edimentary rocks, and the emergence of the conti-ients from the ancient seas.

Another item of the account, founded on geo-ogical investigation, is the period between the drift nd the igneous or plutonic rocks, heretofore known as the primary. This was occupied in the slow deposit of the "sedimentary" or aqueous rocks, having a thickness of two to five miles. Every stone-mason and quarry-man knows that the rocks split in the horizontal much easier than they break in the vertical line. This is owing to the mode of deposition by very gradual subsidence, such as we see when muddy water is allwed too settle and become clear.

The mud at the bottom forms in thin layers or lamina, and when the waters retire as in floods of the Ohio and Mississippi, and the residuary matter becomes dry in the sun, it peels up in thin distinct beds like slate or shale. The secondary shales are the best example of laminated rocks, but the sandstones and conglomerates are stratified, internally, in a less degree. It is this universal feature which distinguishes the aqueous from the igneous rocks, of which there are none in Ohio. The aqueous, or sedimentary, contain fossils, and from these facts we can deduce their great age.

In this State we may examine about 4000 feet in thickness of the fossilliferous, stratified, and laminated rocks. Their order and relative thickness may be seen in the Ohio Reports, extending from the lower silurian, or blue limestone of Cincinnati, up to the top of the coal series.

Now the fact of one bed, or stratum, lying above another in the aqueous rocks, is conclusive as to its relative age.

The upper one must be the most recent. The coal series must be newer than the conglomerate, the conglomerate newer than the fine-grained sandstone, the fine-grained sandstone newer than the black shale, and this newer than the cliff limestone, which is beneath it. Carry the same reasoning into superposition of lamina, and each definable portion of the bed or

;ratum, not thicker sometimes than stout paper, is iore recent than the lamina below it. How many undreds of thousands of these may there be in 4000 ;et?

You will then inquire what is the evidence that iese thin plates of rock were deposited slowly, and ot in a short space of time. If they are closely examined, there will be seen enclosed within the mass ie forms of animals and plants, which retain their orm and minutest markings in all the perfection of fe.

The marine animals, the testacea and the crustacea, iust have lived and grown near where they are now ound, for they are not constructed for travel. The ite of deposition must therefore have been such as to llow a quiet existence, not disturbed by a rapid flux nd reflux of the waters.

In agitated and highly turbid waters, their dead odies would soon be injured, if not destroyed; and t least the delicate markings of these shells would e worn off. The same reasoning may be applied to ie vegetable remains of the coal series. If they ere not deposited in quiet waters, the leaves and fine bres of the stems would certainly be lost; but there not a flower or shrub growing on the face of the lobe more perfect than the coal plants that adorn he roofs of our mines. We know, too, that the deosition now going on in the bottom of our lakes and eas is comparatively slow. When the schooner corn was sunk in Lake Erie, in the fall of 1843, in 8 feet of water, and ten miles from shore, the sediment n her decks at the expiration of two months was bout *one quarter* of an inch, or at the rate of one nd a half inches per year, or twelve and a half feet a century.

If we suppose the sedimentary rocks of the globe

to be three miles in thickness, and I have seen them where they are much thicker, we have 15,840 feet. Allowing a rate of deposition twice as rapid as that upon the deck of the Acorn, or twenty-five feet in a century, and we have 63,200 years for a period occupied in the formation of the sedimentary rocks, and to be added to the period of recession in Niagara and Cuyahoga Falls, making 113,000 years. It may be said that this mode of calculation is very indefinite—that it is founded on speculation, and does not *command* belief.

I think, however, without pursuing the geological evidence farther, every one must admit, that whether we can estimate the existence of the world correctly in this way or not, there are proofs on one side of an antiquity for greater than 6000 years, and on the other no contradictory evidence, even in the Scriptures.

But there is a method of calculation that is liable to no objection, and capable of being made in exact mathematical terms. It is based on a knowledge of the velocity of light. This substance, although it travels very fast, is not, like electricity, endowed with the faculty of instantaneous transit through space. It is eight minutes and thirteen seconds coming to us from the sun. Astronomers were many weary years trying to find out the distance of the fixed stars. Their remoteness was so great, that when viewed from opposite sides of the earth's orbit, that is to say, at opposite extremities of a line, 190,000,000 of miles in length, no "parallax" or angle at the star was observed.

The most they could say was, that if the angle was only *one second*, they should have been able to find it, and that the distance of the star with a parallax of *one second* would be twenty billions (20,000,000,000) of miles, and its light would be about ten years in coming to us. This distance is, to our minds, ıbsolutely incomprehensible, but is less than the nearest fixed stars, and therefore less than the space be-;ween the outside of our system, and the inside of the starry universe beyond.

In comparison with such distances, the orbit of Leverrier's planet becomes a mere line; yet we must ;ax our imaginations to mount still higher, and try ınd grasp the idea of objects many times more remote han *Sirius* or any of the visible stars.

By a process that it would be difficult at this time o make intelligible, Professor Bessel, of Koningsburg, n Prussia, has at last actually solved the problem of he true distance of a fixed star. By years of obser-ation upon a double star in the constellation of the Swan, commonly called "61 cygni," he has found a arallax, by which a mathematical computation, un-rring in its result, gives a distance about three times reater than the least limit just noticed, or 63,000,-00,000 of miles instead of 20,000,000,000—its light eing thirty years in making a journey to us, at the ate of 12,000,000 of miles per minute.

Astronomers take this fixed quantity as a unit, and ith it go on with their examinations of the depths here the remoter stars are seen. This is done by omparing their relative *brightness*, on the supposition ıat they are of equal magnitude, and decrease in rilliancy on account of distance. Applying this rule, ir Wm. Herschel's great telescope brought into view ;ars that were 500 times farther off than 61 cygni, nd thus about 1500 years of our reckoning must pass hile their rays are travelling to our system.

But we can not rest here, although our minds may ;agger at the thought of another forward movement ı space. The power of a telescope is its capacity to ake a body appear to be nearer to us than it appears the eye.

Sirius, or the "Dog Star," might be seen by the naked eye, if removed twelve times farther from us than it is.

If from the last point where it would be thus visible, it was removed again one hundred times farther, and a telescope of the power of 100 applied to it, the observer would bring back Sirius to his vision. The number of stars which we behold in walking out of a clear evening is but few compared to those which the astronomer sees when looking through his telescope, and the higher the power of his instrument, the more numerous do they appear.

Instruments of low power disclose nebula that are imperceptible to the unaided vision, which, on increasing the telescopic power, are found to be not nebula but clusters of stars, so distant that at first they appear like a fleece or cloud of light resting upon the sky. The Milky Way is an instance of a *visible* nebula. Their distance is estimated by the power used in their discovery, supposing the intensity of their light to be the same as other and nearer stars.

Now William Herschel examined nebula, that his telescope did not resolve, or separate into stars, and its power was such that, had they been within a distance from where their light would come to us in 350,000 years, he would have detected the separate stars. Lord Rosse *has resolved some of these nebula* with his 52 feet reflecting telescope, whose known penetrating power is estimated at *ten times* that of Herschel's.

The last-named astronomer supposes that there are luminous bodies within the telescopic range whose rays reach us only after a flight of more than a million of years.

Such is the scale upon which the universe is built. Its remote luminaries are no longer measured by miles, or by diameters of the earth's orbit. These standards

›f measure, though scarcely comprehensible to our enses—*they are so great*—when applied to the disant fixed stars, *are so small* that they do not convey he ideas which astronomers possess and desire to ommunicate.

When they spoke of the comparative distance of he stars among themselves, they state how many years he light of the various stars will occupy in traversing he space between them.

The distance of a star of the first magnitude, like *Sirius*, *Aldebaran*, or *Betelgeuse*, in the shoulder of)rion, they say is *seventeen* (17) years away from the un. Stars of the second magnitude, *thirty* (30) years; nd those of the sixth magnitude, being the most reıote that are visible to the naked eye, are (130) *one undred and thirty years* from the sun.

Those that lie so far out in the ocean of the sky, hat Herschel's great telescope could just discover hem, require 3541 years to make known their existnce to us. We may be overwhelmed by such consideations, our finite minds may be strained and fatigued y them, but we must go farther still.

The sun itself is in motion, bearing along with it ıe whole train of planets, satellites and comets, which onstitute our system.

You will ask where is it travelling to; and moving s it does 33,350,000 miles each year? how long be-)re it will encounter some other system, and produce onfusion there?

Its movement is now towards the constellation of *Tercules*. We should consider here, what I have staэd before, that between the outside of our solar sysэm,—which as now defined, is the planet of Leverrier, nd the nearest fixed star, or sun, or other system,— here is a space which it takes *light*—moving 12,000, 00 of miles every minute of time—*ten years to cross;*

and would require 105,882 years for our earth, going on an errand in space as fast as it does in its own orbit, to reach our nearest celestial neighbor.

The star H in Hercules, to which the sun is directing his course, is *much more distant;* its light instead of being ten years is *forty-six* years in reaching us. The annual rate at which the sun moves, as I have just stated, or 33,350,000 miles, is such that it will require from this time about 1,800,000 years for us to reach Hercules; provided he remains at rest. But not only the constellation "Hercules," but all the constellations and all the stars have a motion, and revolve around a *central point,* which is called the "centre of gravity" of the stellar universe.

This point is near the star Alcyone, one of the *Pleaides,* so far from the earth that its light is 537 years in its flight from there to us.

As a *right line motion* is not known in nature, and since it is well settled that all the planets, satellites, and comets, move in *curves,* we must conclude from analogy, that the sun and all the stars move in curves also, such as circles or ellipses; which curves return to themselves, and constitute an *orbit* like those of the planets, but inconceivably larger. On this hypothesis how long must time flow on, before our sun will complete one journey around Alcyone, and come back to its present position?

Astronomers have computed this period, knowing its present rate of progress, and fix it at 18,200,000 years.!

I have introduced these sublime facts to accustom you to the idea of great periods of time, as well as immense distances in space.

By the side of these vast cycles, what is the historic period of 6000 years? what is a century? what the life of man?

If the ten millions of suns which the telescope liscovers in the heavens have been in existence only ;000 years, our lot is cast in the earliest infancy of he world. For if but 6000 years of the existence of he universe have elapsed, what figures can express he period when its being shall draw to a close? It is easonable to consider that a work of such extent and randeur, should have a length of life in proportion to :s dimensions. That it was not made for display but or some useful and protracted purpose; and if it is rue that the stars, like the planets, have their orbits, nd periods of revolution, it is no doubt true that ley will be permitted to make at least one grand circuit around the universe before they are blotted out.

KETCH OF THE LOCATION, SETTLEMENT, AND PROGRESS OF THE CITY OF CLEVELAND.

[American Pioneer, Vol. II, No. I, January, 1843.]

THIS city is situated on a dry, sandy plain, beveen Lake Erie and the Cuyahoga river; sloping ently towards the lake, from which the water view is xquisitely fine. The plain has an elevation of seven-/-five feet on the lake side, ninety-six at the public luare, and about one hundred and fifteen at the highst point on High street. It appears to have been ocipied in ancient times by the "race of the mounds," a people between them and the present Indian race.

Within the corporate limits of the city, there were traces of two slight works, or lines of earthen embankment, in existence when the white settlement began. One of them is said to have been located on the bluff west of the Light-house; another overlooked the river from a point west of the intersection of Kinsman and Pittsburgh streets. A few low mounds of earth are scattered over the plain.

In the journeyings of the Indians this seemed to be an important point. It is situated at a great southerly bend in the shore of the lake, though not the *most* southerly one, which is near the mouth of Huron river. But the Cuyahoga river enabled their canoes to proceed about thirty-five miles inland, to the "*old portage path;*" from thence a portage of seven miles brought them to the Tuscarawas, a navigable branch of the Muskingum, which communicated with the Ohio. By land, various well-known trails concentrated at the mouth of the Cuyahoga. The opposite banks of the stream seem to have been subject to different jurisdictions from a remote period. The Six Nations and the Wyandot confederacy, nations often engaged in war against each other, both upon land and water, made the Cuyahoga and the "portage path" a part of their boundary.

After the war of the revolution, the British refused to yield possession of the lake country west of this stream, and occupied to its shores until 1790. Their traders had a house in Ohio City, standing north of the Detroit road on the point of the hill near the river when the surveyors first arrived there. By the treaty of Greenville (May, 1795), the western Indians, defeated at Wayne's battle the season previous, relinquished all claim to the lands east of the Cuyahoga, the Portage and the Muskingum, as low down as Fort Laurens, which is near Bolivar in Tuscarawas county

At a council, held at Buffalo in 1796, by General Cleveland and the representatives of the Six Nations, the latter people gave the whites peaceable possession of that part of the Reserve east of this river. It thus remained the line of partition between the white and the red men until July 4, 1805, when the general government extinguished the Indian title to the remainder of the Reserve, by treaty.

When the first county north-west of the Ohio was erected (July 27, 1788), the Cuyahoga was a part of its western boundary, and the lake its northern. After the delivery of the western posts by the English, the *county of Wayne* was set off by the territorial government, with the county seat at Detroit, extending north and west as far as the dominions of the United States; its *eastern* limit was defined to be the course of the Cuyahoga, the Muskingum, and the old portage path. From the time when La Sallé made the voyage of the lakes in the "Griffin," until the abandonment of Canada by the French, in 1763, their traders traversed these regions, and are supposed to have established houses a few miles up the river. After them, the *British Fur Companies* occupied their place, and kept a few small vessels upon the lake. Some years before the settlement here, a schooner, commanded by Captain Thorn, was wrecked a short distance below town, and the crew wintered on shore near the remains of the vessel.

From an early day, the leading Virginia statesmen regarded the mouth of the Cuyahoga as an important commercial position. George Washington, in his journey to the French forts, Venango and Le Bœuf, in 1753, obtained information which led him to consider it as the point of divergence of the future commerce of the lakes, seeking the ocean. Virginia being then regarded as the State through which this trade must pass to the Atlantic, Mr. Jefferson, in his notes upon that State, points out the channel through which it will move to the ocean. He considers the Cuyahoga and Mahoning as navigable, and separated only by a short portage, to be overcome by a canal. Once in the Ohio, produce, in his opinion, might ascend its branches and descend the Potomac to the sea.

In 1795, the Connecticut Land Company was organized at Hartford, Connecticut. On the 5th of September, the fifty-six individuals composing it received a deed from the State of Connecticut of three million of acres, in what was called the Western Reserve. They sent out forty-three surveyors the next year, who were directed to divide that part lying east of the Cuyahoga into townships of five miles square. On the 16th of September, Seth Pease (a brother of the late Judge Pease of the Supreme Court) and Augustus Porter commenced the survey of the "City of Cleveland," as it is called in the minutes. General Moses Cleveland, the agent of the Company, had the honor of furnishing it with a name. This ground had been the source of much controversy between the States, and also the government. Connecticut laid claim to all north of the forty-first degree of north latitude, as far as forty-two degrees two minutes north, and westward to the great South sea, by virtue of a patent from Charles II., king of England. New York procured a patent conflicting with this claim. Pennsylvania and Virginia had also their paper titles upon parts of the same territory. By deed, dated September 13, 1786, Connecticut released all claim to the western lands, excepting and *reserving* New Connecticut, since called "the Reserve." This embraces that part of Ohio north of the forty-first degree, and east of the meridian, one hundred and twenty miles west of the Pennsylvania line. By the deed of cession, Virginia had transferred most of her rights north-west of the Ohio to the United States, March 1st, 1784. Notwithstanding the claims of Connecticut to the Reserve, the United States assumed jurisdiction so far as political sovereignty was concerned, by the ordinance of July 13th, 1787. In 1792, Connecticut made a grant of *half a million* of acres, on the west end of the Reserve, to the sufferers by fire, and therefore called the "fire lands." She still claimed, but did not exercise, jurisdiction of her western province. At length on the 30th of May, 1800, the United States having relinquished all claim to the soil of this tract, the State of Connecticut gave up the right of jurisdiction to the Union at large.

The members of the Land Company, on the same day they received a deed from the State of that part east of the fire lands, conveying to John Morgan, John Caldwell, and Jonathan Brace, all their lands, *in trust*, for specified purposes; and it is through the *quit claim deeds* of these trustees that title to lands in this city and throughout the Company's purchase is derived. The Company paid the State one million two hundred thousand dollars for three million acres, each owner being a tenant in proportion to his stock.

By the close of 1797, the portion east of the Cuyahoga had been laid off into townships. Six of them, including Cleveland, were reserved for private sale, on account of some highly valuable advantages. Four were surveyed into lots of one hundred and sixty acres each, making four hundred in all, to be annexed to the poorer townships, in order to equalize them with *Poland*, the richest of all. The remaining ninety-three were drawn in a lottery, a township for every twelve thousand nine hundred and three dollars twenty-three cents interest, and conveyed by the trustees.

The first drawing took place in February, 1798. In 1836, all the trustees first appointed were living, retained their trust, and executed deeds; Mr. Morgan still survives.

The original plat of the city represents two hundred and twenty lots, seven streets, and four lanes—Superior street, one hundred and thirty-two feet wide; Lake, Huron, and Ohio, each one hundred; and all parallel with the lake were original streets. Water street, Ontario, and Erie, perpendicular to the others, and a public square, thirty-eight by forty rods, were laid out in 1796, and also Mandrake, Union, Vineyard, and Maiden lanes. In November, 1802, Amos Spafford made a re-survey of the streets, altering some and establishing others. Superior lane was laid out by him. The minutes of this survey, in an informal state, were copied into the records of Trumbull county; but, for the most part, no legal record exists of the streets in Cleveland.

With the surveyors, came Mr. Job Stiles and his family, and became the first resident. Judge Kingsbury, now of Newburg in this county, came about the same time, but left his family at Conneaut. Mrs. Stiles was the mother of the first, and Mrs. Kingsbury of the second, white child born on the Reserve. Mr. Stiles left the county in 1798, and Mr. Kingsbury removed to his present farm in the same year. In 1797, Lorenzo Carter became a permanent inhabitant; and soon after him, Nathaniel Doane, who went to Doane's Corners in 1798. Between this time and 1802, Mr. Hally, Mr. Gunn, Stephen Gilbert, Amos Spafford, David Clark, and Samuel Huntingdon, arrived and settled in Cleveland and its vicinity. Mr. Huntingdon afterward removed to Newburg Mills, and thence, in 1807, to Painesville. He was a judge of the Supreme Court of Ohio, and Governor of the State.

Mr. Carter's first cabin stood under the hill, between River street and Mandrake lane, near St. Clair lane. Mr. Clark died in 1806, on the farm across Kingsbury run, on the Pittsburgh road.

In August, 1805, the Cuyahoga was made a port of entry, and John Walworth appointed Collector. His first official duty was the furnishing a clearance to the schooner "Good Intent," which was lost immediately after, near Long Point, crew, vessel, and cargo. He was also made Postmaster, Clerk of the Court, and Recorder; and died in September, 1812.

In July, 1797, the county of Washington was divided, and this place fell within the county of Jefferson, seat of justice at Steubenville. July 10, 1800, the county of Trumbull was established, county seat at Warren; and embracing all of the Reserve. December 31, 1805, a new division took place, which left Cleveland in the county of Geauga; organized March 1st, 1806. On the 10th of February, 1808, the county of Cuyahoga was erected; organized May 1st, 1810, with Cleveland as the county seat. The first Court of Common Pleas was holden June 5, 1810. Of this Court, Benjamin Ruggles was presiding judge, Major Nathan Perry, Timothy Doane and Augustus Gilbert, associates. Of the first grand jury, James Kingsbury was foreman; a place to which, by long usage, he seems to have acquired a kind of prescriptive right. The Supreme Court held its first sitting on the 13th of August, 1810; judges, William W. Price and Ethan Allen Brown. At the April term, 1812, an Indian of the Chippeway tribe, by the name of John-O-Mic, was indicted for the murder of Daniel Buel, a white man, at Pipe creek, near Sandusky city. And also as an accomplice with *Lemo*, an Indian of the same tribe, in the murder of Michael Gibbs, at the same place. It appears that O-Mic killed Buel with his tomahawk, and Lemo shot Gibbs with a pistol at the same time. Their object was merely the plunder of a few articles of goods and clothing. Lemo retreated to his tribe, at Cedar Point, for protection. The officers of justice pursued him, and, arriving at the camp, found his body stretched upon the ground without life. Through fear of the United States he had been bound, for the purpose of being delivered up to justice. To avoid this, he rolled himself to a tree, against which stood a loaded gun. Though pinioned with his arms behind, he contrived to place the muzzle to his throat, and discharged the piece with his toe. O-Mic was convicted and sentenced to be hung on the 26th of June, 1812. He had been confined since the arrest, by a chain and staple, to the floor of Mr. Carter's ball-room, in the old red house, formerly standing beetween Water street and Union lane, near Superior lane, and had grown fat and strong. At the hour of execution, he objected to going upon the scaffold; this difficulty was removed, however, by the promise of a pint of whiskey, which he swallowed, and took his departure for the land of the great spirit. This event was witnessed by large numbers of citizens, from this and adjoining counties, at the centre of the public square. They were assembled with arms, under apprehensions of an attempt, on the part of the Indians, to rescue O-Mic.

The declaration of war, in June of the same year, placed our city in an important and dangerous position. In 1813 it became a depot of supplies and rendezvous for troops destined for operations farther west. A small stockade was erected at the foot of Ontario street on the bank of the lake, the outlines of which are still visible. A permanent garrison of infantry, under Major Jessup, now quarter-master general of the United States army, occupied the place.

Although the British vessels appeared frequently off the town, no attempts at landing were made here. The greatest alarm was occasioned by the approach of boats containing the prisoners surrendered by Hull at Detroit, who were at first mistaken for British and Indian troops advancing to storm the place.

At length peace came, and the alarms of war being past, the occasion was thought to be worthy of celebration, by libations of whiskey, and the discharge of cannon. During the performance, the present sexton of the graveyard, Abram Hickox, carried the powder in an open tin pail upon his arm; to fire the gun, another carried a stick with fire at the end, kept alive by swinging it through the air. Amid the general excitement, a spark from this brand found its way into uncle Abram's powder about the time the gun was discharged. The body of the worthy sexton was seen to rise through the air as high as the eaves of the house, and returned to the earth blackened and destitute of clothing. According to his own vociferations he was already dead, but the bystanders thought differently; and carried him to a room, where it was found that he was not dangerously wounded.

By an act of the legislature, dated December 23, 1814, the village was incorporated, with limited powers, administered by a President, Recorder, and three Trustees. In March, 1836, a more extensive charter was obtained, with full municipal authority, exercised by a council, composed of the Mayor, three Aldermen, and of Councilmen, three from each ward.

The Ohio Canal was commenced in 1825, and connected with the Cuyahoga near its mouth. By means of the experiments at Buffalo and Erie, it was now settled that artificial harbors might be made in the still water at the mouths of lake streams. Here a harbor was indispensable to connect the lake and canal navigation. Mr. A Kelly, then one of the canal commissioners of Ohio, considered the construction of one at this point as practicable, and reported to the legislature that five thousand dollars would be sufficient for the object. Upon the importunity of citizens, backed by the authority of this report, Congress, with great reluctance, appropriated five thousand dollars in the winter of 1824–5. It was confided to A. W. Walworth, Esq., collector of the port, without surveys or even instructions as to the manner of disbursement. Upon consultation it was determined to start a pier from the shore, outward into the lake, nearly at right angles, being in the direction north thirty-two degrees west, beginning about forty rods east of the point where the river then discharged itself. Here, as at all other streams on the lakes, there was a channel, at times, at others not. To those acquainted with the action of waves which strike obliquely upon a shore, it is known that a lateral current is created of much force. It is thus that masses of sand travel up and down the coast according to the prevalence of the wind. Here, the north-*eastern* winds predominate, and therefore the eastern bank of the stream has a tendency to prolong itself *westward* at the mouth, and cause the river to enter the lake obliquely. The object of this single pier was to catch the drift from the east and to prevent it from filling the channel. In the natural state, it was very seldom that a schooner of fifty tons could enter without lighterage, and many times a common row boat would touch. The pier was extended into the water six hundred feet, with the five thousand dollars first appropriated. As usual, the majority predicted unfavorably to the success of the project; even some of those concerned had doubts. The pier being now finished, no particular change in the capacity of the channel was observed. Congress ıd given all that was asked to complete a harbor; it ıd been disbursed and no apparent benefit had fol- wed. The faithless had the satisfaction of seeing the outh of the river more often clogged with sand than ›retofore, and, in one instance, a bar of dry land ex- nded entirely across its entrance. This, they said, ıs the port of Cuyahoga.

But faith still remained to some of the citizens, ıd a meeting was called late in the fall of 1825, eeches made, and two committees of five appointed. ıe of them was charged with the selection of an ent to proceed to Washington and solicit more aid ›m Congress; the other with the duty of raising e hundred and fifty dollars for his expenses. A. . Walworth, Esq., was engaged to visit the federal ;y, which he reached in January, 1826. Prospects ›re by no means encouraging. Cleveland at that ne presented itself to the minds of the eastern mem- rs, as a straggling village, on the frontiers of civili- tion. Its relations were not understood; its few ıabitants shaking with the ague, and only thirty or :ty arrivals of vessels in a year! It is, therefore, t strange that amid the moves and counter-moves the great capital, such a city should lack influence, its claims fail to receive attention. Mr. Walworth ınd that personal statements would not be received the committee (on commerce), and that Mr. Cam- ›ling, of New York, was decidedly opposed to grant- ; assistance to any of the western ports. Governor mlinson, of Connecticut, also a member of the com- ttee, took an opposite view, and entered warmly o the project. Mr. Whittlesey, of Ohio, labored ›iduously and eventually with success. But it was torious, that with the five thousand dollars, which s all that had been at first demanded to complete › work, no certain benefit had been derived. On the other side, the "Washington," the "Erie," a: the "American Eagle," had been stranded in sig of the proposed harbor, and property to the amou of thirty thousand dollars lost to the owners. It w evident that commerce must cease or harbors be bui If our government would not make them, the shi ping interest would, in the end, be compelled to do

In 1818, a steamboat made its appearance on the waters, called the "Walk-in-the-water." In anticip tion of the construction of artificial harbors along t shore, as had been already done at Buffalo, these cra had increased in number. The Walk-in-the-wat unable to make shelter, was lost in a storm in the fa of 1821. The steamboats "Superior," "Henry Clay "Enterprise," "Pioneer," and others, made the pa sage of the lakes in fear, and without regularit Passengers could be landed as well at one point of t coast as another, but only when calm weather preva ed. There were at times exceptions to this rule, Erie and Sandusky; but persons destined for th place were carried by twice, or even thrice, witho being able to land.

Finally, the rule of the committee on commerce r quiring statements in writing was rescinded and Messr Whittlesey and Walworth appeared before them in pe son. Shortly after, they reported in favor of gran ing ten thousand dollars in addition to the sum of fi thousand dollars heretofore expended, being in full tl amount supposed to be necessary to completethe wor Ten thousand dollars was finally appropriated by la but not in time for use in 1826. In the spring of 182 Major T. W. Maurice, of the corps of engineers, a rived and examined the premises. He made a su vey, and reported a plan for the work, of which t present piers are the result. It was determin that the channel, instead of being formed *west* of t

er, as constructed by Walworth, should be form-
l on the *east* side of it, between this and another
rown out nearly parallel and about two hundred feet
stant. A few piles had been driven in the river to
rn the current across, as originally designed, but with-
t effect. Major Maurice ordered a dam to be built
rectly across the river, opposite the south end of the
st pier. The construction of this barrier occupied
e season. An opening had been left for the pas-
ge of such craft as were able to enter the river, un-
late in the fall, when it became necessary to test
principle of the work, close it up, and direct the
ter of the river across the neck of land to the lake.
many of the lake captains and proprietors of ves-
s, the idea of forming a channel where nature had
led appeared to be absurd, and the expenditure of
ney for such a purpose equivalent to throwing it
ay. They saw the natural harbor, poor as it was,
out to be closed. The prospect of forming another
that means they considered as nothing at all. It
peared to them that their natural rights of naviga-
n were to be taken away by an audacious display
federal authority, and from harmless abuse of this
rk and the workmen, they passed to more serious
, and some to threats of violence. The gap in
dam was closed. The schooner "Lake Serpent,"
tain Foster, entered the river, and was soon after
t in below the dam by a bar at the mouth. The
tain dug a way out for her, and set sail in a rage,
laring that he only required a lease of life until
y should have formed a harbor upon the plan now
ler experiment. The moment of trial had now
ne, and the fall rains began to raise the Cuyahoga,
was expected. Men with saws and axes, and oxen
h chains and scrapers, were put to work to make a
e across the neck and lead the current in its des-

tined course. On the morning of the 22d of Octobe: after the work had been prosecuted many hours ami the cold storms of the season, and while the dam ap peared about to yield to the pressure of the flood, th water began to flow direct to the lake, through th sand and flood-wood, east of the first pier. By th time this flood subsided there was *two* feet water in th new channel, which was continually enlarging. Whe the "Lake Serpent" returned from her cruise, she foun the new harbor capacious enough to admit her ke without difficulty. The old channel, through whic water no longer passed, was soon sealed up with san In the spring of 1828, the eastern pier was com menced, beginning at the water's edge. The wo has progressed steadily until 1840, at a cost of $77 550.

The western pier was soon connected with th dam, and the eastern continued in a southerly dire tion across the neck to the river. Both have bee carried outward into the lake, making twelve hundre feet in length. About four hundred feet of the eas ern pier, having partially decayed, has been replace by heavy cut stone masonry, having a foundation u on the original work, the piles and stone removed b low the surface of low water in the lake. It is pr posed to renew the whole in this manner, which w leave it in an imperishable state, a monument (the liberality, grandeur, and utility of the nation Union when, perhaps, that institution itself shall l known only in the history of things that were.

In the year 1830, a light-house was built on lan its base seventy-five and its light one hundred an thirty-five feet above the lake. Since that time beacon-light has been erected at the extremity of th eastern pier, about forty feet in height, and, in a mea ure, superseding the light on Water street.

The elevation of the surface of the lake fluctuates, ı the extreme, about five and a half feet, and conse-uently the expression of a height above its surface true only for the precise time when it was taken. 'he rise and depression of each year is from ten to ghteen inches, being at high water late in the spring ıd low water late in the fall or early in winter. he general period of high water may be half a cen-ıry, or even more. In June, 1838, the rise had ached its greatest known height at this place. The ırface on the 25th of that month corresponded with ıe lower face of the *second* course of masonry, from ıe top of the east pier at its southern extremity. t this time it ranges about two and a half below that ark. The height of water in the lakes, like that in vamps and rivers in general, is controlled by the ıaracter of the seasons, and needs no other explana-ɔn. It is probable that the settlement of the coun-y has a tendency to increase the volume of water scharged to these reservoirs above that of former nes, by causing a more rapid drainage. If this is ue, we may expect a higher state in future than has en experienced for the past fifty years, varying di-ctly with the *rain guage* of the lake country and e mean temperature of the atmosphere.*

The anticipated importance of a harbor at the ıyahoga has been fully realized. In 1840, there ssed the piers, inward bound, one thousand three ındred and forty-four vessels, and one thousand ıd twenty steamboats. *Seven* steamboats, fifty-four hooners, and two brigs, with an aggregate tonnage nine thousand five hundred and four, belonged to is port. The tonnage of steamboats entering was ree hundred and fifty-seven thousand, of vessels one

* See statistics of levels at the end of this article.

hundred and twenty thousand nine hundred and sixty The exports for that year were estimated at *five mil lion dollars;* of which two million one hundred thous and bushels of wheat, and five hundred thousand bar rels of flour, formed a great proportion.

New York, by way of Buffalo and Albany, is sev en hundred and six miles; by the New York and Eri Railroad, six hundred. To Philadelphia, by canal, si hundred and nine; by road, through Pittsburgh an Chambersburgh, four hundred and thirty-two. T Baltimore, by Cumberland and the Point of Rock four hundred and fifty-three. We thus perceive ho much reason General Washington and Mr. Jefferso had to suppose that produce would seek the Atlanti ports at the south, rather than at the north, so far a the *distances* are concerned. Furthermore, it wa probably known to them, that the lake is open for na igation from Cleveland west; earlier in the sprin than it is farther east. From 1829 to 1837, inclusiv the average difference between Cleveland and Buffal in this respect, was *thirty-three days*—the greate difference *sixty-five days*, the least three. This is o casioned not so much by the climate as other cause In winter, the lake is never wholly closed by ice. A this place its width is about sixty miles. Its surfa is frozen to a distance of twenty miles from t shore, but probably not much further; for when t south wind blows long off shore, the ice is seen move away in a body, and frequently passes beyo the range of view from the top of the light-hous To effect this, it must proceed twenty-five or thir miles from the American side towards the Canadia If it floats thirty miles from shore, there are thir miles occupied by ice formed along both shores wh driven together by the winds. Now when the i breaks up in the spring, the westerly winds and t natural flow of the water towards Niagara force it towards the outlet at Buffalo, thus the vacancy in the central part of the lake affords room for much or all of the ice of the western half. Here it remains, choking up the lake, as far as Ashtabula, and sometimes to Fairport, until dissolved by the influence of the sun, or broken up by the winds, and borne by the current over the cataract of Niagara.

Since the settlement of the town, the waters of the lake have encroached upon its site at the rate of *forty rods* or one-eighth of a mile in a century. The distance from the water's edge to the bend of the Cuyahoga, along Ontario street and the public square, is but little over half a mile; and therefore it will be seen that if this advance should go on unchecked, the Cuyahoga might discharge itself between Seneca and Ontario streets within *five hundred years*. If the slope of the ground, which is already gone, was the same as that which remains and upon which the town is built, it would descend to the lake level within *two miles*. This distance would consequently be the measure of the recession of the shore at this place, and fix the ancient mouth of the Cuyahoga within two miles of the present shore. It is almost certain, that when the French missionaries rowed along these shores, in 1673, the bluff in Ohio City and that in Cleveland were united, and the river passed into the lake at the old mouth one mile westward.

An artificial protection against the inroads of the lake was made along a part of the front of the city, from Ontario to Seneca streets, in 1841. It consists of a breakwater of piles and stone, and a grade of the bank to an angle of quiescence, which is about fourteen degrees. If it shall prove to be an effectual barrier, the work will doubtless be extended so as to cover the town.

The population of Cleveland, in 1840, was 6,071 The largest church in this city is the Presbyterian there is one Episcopal, one Methodist Episcopal, an one Methodist Protestant church; also one Congre gational, one Dutch Reformed church, and a syna gogue of Jews.

MEASUREMENTS UPON THE RISE AND FALL OF THE SURFACE OF THE NORTHERN LAKES.

[True Democrat, Cleveland, April 23, 1851.]

MESSRS. EDITORS: As far as I know, there are n precise data by which to determine the fluctuation in the waters of our Northern Lakes, previous to th year 1819. What is known of the subject prior t that time rests upon tradition or vague recollection and not upon recorded measurements.

Shortly after the war of 1812, Colonel Henr Whiting, a Quartermaster of the United States Army stationed at Detroit, made regular observations in th Detroit River, and from 1819, for several years, h made and preserved measurements of the stage of th water.

From 1828, for some years, A. E. Hathon, Esq. a civil engineer, connected with the old water-works a Detroit, kept an account of the height of the water a that place; and in 1830 the late Dr. Houghton bega to make his observations.

The geological reports of Michigan, for 1839 an 1841, contain the substance of what was known re specting lakes Huron and Superior at that time, pre pared by Dr. Houghton, the principal geologist, an by S. W. Higgins, Esq., the topographer of the sur vey.

When the Ohio survey was in progress, some information was collected, principally traditionary, which may be seen in the 2d Report, pages 50, 51 and 52.

The first daily water tables kept on Lake Erie, so far as I am informed, were made in 1838, at my request, by Mr. George C. Davies, under the direction of A. Walworth, Esq., disbursing agent for the government works at Cleveland.

The obseravtions of Mr. Davies were made almost hourly (during the day), from the 18th of August to the 1st of December, upon a float, placed in a slip near the office, where the water was not disturbed by the waves. While they were continued they furnish a faithful and perfect representation of the lake surface.

In 1844, when Colonel T. B. W. Stockton took charge of the construction of the government works at the harbor, he caused a meteorological table to be kept, which is very complete and reliable.

From August, 1845, to August, 1846, inclusive, Colonel Stockton added a water table. The observations were taken at 6 A. M., at noon, and 6 P. M., by Mr. George Tiebout, of this city, which by the kindness of Mr. D. P. Rhodes, late disbursing agent, I have in my possession.

Since 1838, I have occasionally taken the height of the water, but not with any approach to that regularity and system which is necessary to give value to such observations.

The most extended measurements in point of time have been made at Black Rock, N. Y.; but at what time they commenced, or whether they are still continued, I cannot say, not having as yet succeeded in procuring anything more than occasional published reports from the Buffalo papers. They are taken by the engineers of the Erie Canal, which is supplied by water from Lake Erie; and the depth of water on the mitre sill of the guard-lock at Black Rock represents the comparative elevation of the lake at that place. It is probable that this record was commenced soon after the completion of the canal in 1825, and is still continued.

From these sources I have formed an abstract of the rise and fall of this and other lakes since 1796, as complete as the present state of information in my possession will allow, embracing the traditionary as well as the authentic, which is introduced into a report recently made to the government on the geology of the public lands of the upper peninsula of Michigan.

I was engaged in making an abstract of the water table of Colonel Stockton for publication in some of our city papers, by condensing the daily observations into the mean expression for the week, when the request, contained in your editorial of the 14th instant, fell under my notice, where you wish to have some theory or explanation given of the reputed rise of seven years' duration and a succeeding fall or depression of seven years. The table shows no such regular rise and fall; and the only *theory* or *mystery* there is connected with the subject, is that of the *fluctuations of the seasons*, operating upon these *great ponds* as upon all reservoirs of water. When rains and cloudy and cool weather prevail over the lake country, in due time the lakes and rivers *rise*, and when the reverse happens they sink away.

The results of the rain guage and the observations of lake levels go *exactly together*. There is also an annual rise and fall owing to the seasons. But the subject is a long one, and at present I can do no more than promise to furnish you at some future time some figures, showing the various stages of the lake since 1819, by recorded observations, from which every one may draw their own conclusions.

Lake Levels.—1845 & 1846.

Abstract of the Water Table in Colonel T. B. W. Stockton's Meteorological Record, kept at the Harbor Office in Cleveland.

The *line* of reference is the high water line of June, 1838, being the highest known stage, and two feet below the top of the main wall of the East Pier, 400 feet from the south end and near the south end of the parapet wall, which is four feet higher than the main wall. The figures show the *depression* of the water surface below June, 1838, giving the *mean* of each week's observations three times a day.

The three lines of figures under the head of the month represent the mean height at 6 A. M., at noon, and 6 P. M., daily, for each week.

1st WEEK.	2d WEEK.	3d WEEK.	4th WEEK.	MONTHLY MEAN.
		August, 1845.		
wanting.	wanting.	2 ft. 9 in.	2 ft. 8,9 in.	
		2 ft. 6,7 in.	2 ft. 8,9 in.	
		2 ft. 7,6 in.	2 ft. 9,1 in.	
		September, 1845.		
2 ft. 11,4 in.	2 ft. 11 in.	3 ft. 0,8 in.	3 ft. 1,7 in.	3 ft. 0, 3 in.
2 ft. 9,1 in.	2 ft. 11,3 in.	3 ft. 0,1 in.	3 ft. 0,6 in.	2 ft. 11, 2 in.
2 ft. 8,8 in.	2 ft. 10,7 in.	2 ft. 11 in.	3 ft. 0,9 in.	2 ft. 10, 8 in.
		October, 1845.		
3 ft. 2,3 in.	3 ft. 2,3 in.	3 ft. 3,0 in.	3 ft. 3,7 in.	3 ft. 2,75 in.
3 ft. 1,7 in.	3 ft. 1,3 in.	3 ft. 0,3 in.	3 ft. 3,9 in.	3 ft. 1, 8 in.
3 ft. 0,3 in.	3 ft. 0,9 in.	3 ft. 2,1 in.	3 ft. 3,8 in.	3 ft. 1, 8 in.
		November, 1845.		
3 ft. 2,0 in.	3 ft. 5,7 in.	3 ft. 8,4 in.	3 ft. 8,6 in.	3 ft. 6,18 in.
3 ft. 2,1 in.	3 ft. 6,1 in.	3 ft. 8,0 in.	3 ft. 7,4 in.	3 ft. 5, 9 in.
3 ft. 2,8 in.	3 ft. 4,3 in.	3 ft. 6,7 in.	3 ft. 5,6 in.	3 ft. 4,85 in.
		December, 1845.		
3 ft. 7,6 in.	3 ft. 7,8 in.	3 ft. 10,5 in.	4 ft. 2,0 in.	3 ft. 9,9 in.
3 ft. 8,1 in.	3 ft. 7,8 in.	3 ft. 11,4 in.	4 ft. 0,0 in.	3 ft. 9,9 in.
3 ft. 5,0 in.	3 ft. 8,1 in.	3 ft. 10,0 in.	4 ft. 0,0 in.	3 ft. 8,8 in.
		January, 1846.		
4 ft. 0,0 in.	3 ft. 11,7 in.	4 ft. 1,3 in.	4 ft. 1,8 in.	4 ft. 0,7 in.
4 ft. 0,8 in.	4 ft. 0,4 in.	4 ft. 0,5 in.	4 ft. 2,0 in.	4 ft. 0,9 in.
3 ft. 11,5 in.	3 ft. 11,1 in.	4 ft. 0,0 in.	4 ft. 2,7 in.	4 ft. 0,3 in.
		February, 1846.		
4 ft. 2,1 in.	4 ft. 4,2 in.	4 ft. 6,1 in.	4 ft. 6,3 in.	4 ft. 4,7 in.
4 ft. 2,5 in.	4 ft. 4,2 in.	4 ft. 6,4 in.	4 ft. 6,3 in.	4 ft. 4,8 in.
4 ft. 2,3 in.	4 ft. 4,5 in.	4 ft. 5,7 in.	4 ft. 7,1 in.	4 ft. 4,9 in.

March, 1846.				
4 ft. 8,8 in.	4 ft. 8,0 in.	4 ft. 6,5 in.	4 ft. 2,7 in.	4 ft. 6,5 i
4 ft. 6,3 in.	4 ft. 8,1 in.	4 ft. 4,7 in.	4 ft. 2,5 in.	4 ft. 5,4 i
4 ft. 5,3 in.	4 ft. 8,5 in.	4 ft. 6,3 in.	4 ft. 2,7 in.	4 ft. 5,7 i
April, 1846.				
4 ft. 1,5 in.	4 ft. 1,3 in.	4 ft. 0.3 in.	4 ft. 1,3 in.	4 ft. 1,1 i
4 ft. 2,3 in.	4 ft. 0,3 in.	3 ft. 11,8 in.	4 ft. 0,3 in.	4 ft. 0,6 i
4 ft. 1,3 in.	4 ft. 3,7 in.	3 ft. 11,3 in.	4 ft. 3,7 in.	4 ft. 2,0 i
May, 1846.				
3 ft. 6,3 in.	3 ft. 3,1 in.	2 ft. 9,0 in.	3 ft. 0,0 in.	3 ft. 1,2 i
3 ft. 6,6 in.	3 ft. 3,3 in.	2 ft. 10,0 in.	2 ft. 11,2 in.	3 ft. 2,0 i
3 ft. 5,6 in.	3 ft. 3,5 in.	2 ft. 10,9 in.	2 ft. 10,9 in.	3 ft. 1,7 i
June, 1846.				
2 ft. 8,5 in.	2 ft. 10,5 in.	2 ft. 10,5 in.	2 ft. 10,5 in.	2 ft. 10, 4 i
2 ft. 11,4 in.	2 ft. 9.0 in.	2 ft. 10,1 in.	2 ft. 11,5 in.	2 ft. 10, 5 i
2 ft. 10,2 in.	2 ft. 10,7 in.	2 ft. 8,1 in.	2 ft. 11,2 in.	2 ft. 10, i
July, 1846.				
2 ft. 7,0 in.	2 ft. 8.8 in.	3 ft. 0,0 in.	3 ft. 2,4 in.	2 ft. 10,5 i
2 ft. 8,8 in.	2 ft. 9,8 in.	2 ft. 11,3 in.	3 ft. 2,0 in.	2 ft. 10,9 i
2 ft. 9,4 in.	2 ft. 10,8 in.	2 ft. 8,8 in.	3 ft. 2,0 in.	2 ft. 10,7 i
August, 1846.				
3 ft. 1,8 in.	3 ft. 1,7 in.	3 ft. 0,9 in.	3 ft. 1,5 in.	3 ft. 1,5 i
3 ft. 1,1 in.	3 ft. 1,7 in.	2 ft. 6.7 in.	3 ft. 0,8 in.	2 ft. 11,6 i
3 ft. 1,7 in.	3 ft. 1.4 in.	2 ft. 10,9 in.	3 ft. 2,7 in.	3 ft. 1,2 i

The lowest stage of 1846 was in the month of March.................. 4 ft. 5,8 i
The highest stage of 1846 was in the month of June................ 2 ft. 10,1 i

Greatest difference of the monthly mean for 1846...................... 1 ft. 7,7 i

Beginning at December, 1845, and extending April, 1846, the general surface varied only about inches. The month of September, 1845 (2 ft. 11 in.), is but little below the month of June, 1846, or ft. 10,1 in. The last two weeks of August, 1845, sh a higher stage than any part of 1846, or 2 ft. 8,7 i

This table illustrates very well the *regular annu* rise and fall which has always been noticed by tho who keep these lake registers. The recession fro high to low is as systematic and certain as the season but the month in which high water of one year occu may not be the same as another, and the same low water. In general the lowest ebb occurs the latter part of winter, when the supply of wat has been long withheld by the presence of snow a frost. The flood of the season happens generally

June, when the streams have discharged the surplus of the spring months.

The following average of each month in 1846 shows the perfect regularity of the *annual* elevation and depression. If we had as perfect a table for *several consecutive years* as we have for this, a *yearly abstract* would show the *general* elevation and depression in the same way.

September,	1845	average,	2	feet	11,4	inches.
October,	"	"	3	"	2,4	"
November,	"	"	3	"	5,6	"
December,	"	"	3	"	9,5	"
January,	1846	"	4	"	0,6	"
February,	"	"	4	"	4,8	"
March,	"	"	4	"	5,8	"
April,	"	"	4	"	1,2	"
May,	"	"	3	"	1,6	"
June,	"	"	2	"	10,1	"
July,	"	"	2	"	10,7	"
August,	"	"	3	"	00,7	"

A Table showing the highest and lowest stage of water in Lake Erie for each month, *from August,* 1845, *to August,* 1846, *inclusive, drawn from Colonel T. B. W. Stockton's Meteorological Record, as kept at the government works on the Harbor at Cleveland. Zero of reference, the high water of June,* 1838.

TIME.	WINDS, ETC.	HIGH WATER	LOW WATER	GREATEST MONTHLY difference
August, 1845.		*below* °	*below* °	
Several times	wind southerly	2 ft. 6		
6 A. M. 14th	high south wind		3 ft. 4	0 ft.10
September.				
A. M. of 5th, and M. of 24th	heavy N. W	2 ft. 6		
Noon of 26th	light S. W.		3 ft. 7	1 ft. 1
October.				
3d, 6 P. M.	high N. E.	2 ft. 8		
24th, 6 A. M	light S		3 ft.10	1 ft. 2
November.				
Eve. of 2d and A. M. of 3d	gale N. E.	3 ft. 8		
Morn. of 19, at 4 A. M., water settled to 6 ft. 2 in. B. for a short time	strong S. W.		4 ft. 8	2 ft. 0
December.				
1st morning	gale N. E.	2 ft. 6		
16th morning	breeze north	2 ft. 6		2 ft. 5
27th noon	fresh breeze S. W.		4 ft.11	
January, 1846.				
31st morning	high N. E.	3 ft. 5		
10th noon	fresh S. W		4 ft. 9	
18th morning	fresh N.		4 ft. 10	1 ft. 5

Table Continued.

TIME.	WINDS, ETC.	HIGH WATER	LOW WATER	GREATEST MONTHLY difference
February.				
10th, morning	fresh N. W	3 ft. 10		
21st, morning	high S. W		5 ft. 1	1 ft. 3
March.				
30th, morning	fresh W.	4 ft. 0		
7th, noon	light W		5 ft. 0	1 ft. 0
April.				
25th, noon	heavy gale N. E.	3 ft. 6		
27th noon	do. do. do.	3 ft. 6		
30th, evening	do. do. do.	3 ft. 6		
11th, evening	heavy W		5 ft. 0	1 ft. 6
At 3 P. M., 5 ft. 6 in. B				
May.				
18th evening and 19th morning.	gale N	2 ft. 9		
3d, noon	light S		3 ft. 9	
5th and 6th noon	light N. E		3 ft. 9	1 ft. 0
June.				
2d morning	fresh N. W	2 ft. 6		
20th, evening	high N	2 ft. 6		
21st, morning	gale N. E	2 ft. 6		
30th, morning	fresh south	2 ft. 6		
14th, morning	breeze N		3 ft. 2	0 ft. 8
July.			3 ft. 4	
11th, 6 A. M	light south	2 ft. 6		
21st, 6 A. M	do. do		3 ft. 4	0 ft. 10
August.				
18th, noon [average of the 18th, 2 ft. 0,6 in.]	breeze N	1 ft. 10		
30th, 6 P. M.	breeze N. E.		4 ft. 9	2 ft. 11
At noon 3 ft. 2				

It is apparent from this exhibit that north-east winds and gales prevail over those of any other direction, and produce most uniform effects in heaping or *raising* the waters at this place; and also that south-west winds have the contrary effect, and drive out and depress the water of the lake.

Of twenty-one cases of extreme *high* water above given, eight are due to north-east winds, three to south winds, and three to north-west.

Of sixteen instances of low water, two are connected with north-east winds, two with westerly, and five with south-west winds.

High water occurred oftener in the morning than

ıt noon or evening, and the same of low water, which s a consequence of the changes that occur in the and and sea breezes near the commencement and close of the day. The water has been high in the norning (6 A. M.) ten times, in the evening five, at ıoon twice.

It has been low in the morning seven times, at noon six, in the evening twice.

The greatest fluctuations take place in the morn-ng; of which the most extreme case on these records occurred on the morning of November 19, 1845, after two days' light breeze south, changing to high south-west winds. The water then fell 20 inches below the bottom of the stone wall, or to 6 feet 2 inches.

It is apparent, from the column of "differences" in the preceding table, how little reliance can be placed on a *single* measurement, and how necessary it is to have a daily register in order to arrive at the mean surface of the lake. Winds not felt at this point may affect the stage of water materially.

LAKE LEVELS.—GENERAL ABSTRACT.

MESSRS. EDITORS: I am unable to present you at this time a perfect table of the reported stages of water in Lake Erie, on account of the difficulty of reducing observations taken at different points of the lake, at different times, and referred to different *lines* of level, to *one standard* of zero.

At Detroit, the *low water* of August, 1819, is taken as the line of reference, counting *upward*.

At Cleveland, we use the *high water line* of June, 1838, as zero, and count downwards.

At Black Rock they reckon from the mitre sill of the guard-lock, and give the depth of water at the lock. There is another zero used at Buffalo harbor, which is 5 feet 4 inches below the high water of June, 1838.

If we had consecutive observations at all these places, even for *one month*, a pretty close reduction could be made of the different zeros, or standard lines, to each other; but as yet I have found only part of a month on which reliance can be placed; and as the lake is never precisely level, there might be a difference or error in the comparison *greater* than the mean annual change of surface.

I give you, however, a few tables drawn from readings at the above places, which serve as well to show the *fluctuations* at these places as if the absolute average rise and fall of all parts of the lake was known, and reduced to *one standard*. This is done by taking the stage of water in the *same month*, through a series of years.

Different observers do not agree as to the precise time of either the highest or lowest general state of the water. Here, it was unquestionably the *highest* about the 25th of June, 1838, when it was 2 feet 5 inches below the surface of the main wall at the south end of the east pier; 2 feet below the same wall 400 feet from the south end; 6 feet 4 inches below the coping of the lock at the Cuyahoga river, at Merwin street (on the west wing wall); 3 feet 88-100 below the water table of the Commercial House, corner of Main and river streets, Ohio City; and was also on a level with the heads of the piles on the river, at the Morocco Factory on the flats.

These marks and benches were established by Mr. Howe, the resident engineer of this division of the Ohio canal, by General Ahaz Merchant, by Colonel T. B. W. Stockton, and by myself, independently of each other, and will probably some of them remain for future reference, through all time.

At Detroit, and Black Rock, the highest stage is stated to have been in August, and not in June, 1838; but at Buffalo it is given in the month of June. Some observers place the lowest water in August, and some in the fall of 1819; others in Autumn, 1820; and at Buffalo there is a tradition that in 1810 it was still lower.

COMPARISON

Of the stage of water in the month of June, for several years at Cleveland.

1832, below, June, 1838	2 ft. 6	inches.
1833, " " "	2 ft. 8	"
1834, " " "	2 ft. 7	"
1835, " " "	2 ft. 10	"
1837, " " "	2 ft. 7	"
1838, " " "	0 ft. 0	"
(mean of the month,)	2 ft. 10	"
1850, " " " (one measurement)	3 ft. 0	"

COMPARISON

For the month of June, at Detroit.

1828, below, August, 1838	3 ft. 2	inches.
1830, " " "	4 ft. 2	"
1836, " " "	2 ft. 4	"
1837, " " "	1 ft. 14	"
1840, " " " (one measurement)	2 ft. 4	"

COMPARISON

For the month of August, at Black Rock.

1828, above, August, 1820	2 ft. 6	inches.
1830, " " "	2 ft. 6	"
1836, " " "	3 ft. 6	"
1837, " " "	4 ft. 0	"
1838, " " "	5 ft. 1	"
1839, " " "	3 ft. 10	"
1840, " " "	2 ft. 7	"
1841, " " "	0 ft. 9	"

In these abstracts we have all of the years from 1828 to 1841, except one, and when the measurements at the several places can be reduced to one common line or plane of reference, a statement of the absolute elevation and depression of the lake for those thirteen years can be made out.

Owing to the form of the shores which approacl each other at Buffalo, the greatest temporary fluctua- tions take place there. The greatest measured loca range, as given by a correspondent of the Buffal *Commercial Advertiser*, is fifteen feet six inches; th highest stage occurred during a furious down-lak storm, October 18 and 19, 1844; the lowest, unde an up-lake gale, April 18, 1848.

Cleveland being near the broadest part of the lake and towards the middle of its length, is less affecte of than places at either extremity, but here a differenc more than three feet has been observed in the same day

It should not be forgotten that the preceding ta ble is not *absolute*, but only *comparative*, showing th rise and fall of corresponding months in differen years. June is generally a *high* month, and Februa ry a *low* one; but this is not invariably the case; s that the average of years can not be had by knowin how the water stood in the same months. To obtai this, we want all the months of the year, and *dail* observations for each month.

From the above it will be seen that the greates well ascertained *difference of level* at Cleveland i between June, 1838, and June, 1846, 2 feet 10 inches at Detroit, from August, 1828, to August, 1837, 1 foot 3 inches; at Black Rock, between August, 1838 and August, 1841, 4 feet 4 inches; at Buffalo (b single measurement), between May, 1840, and May 1846, 1 foot 9 inches; results do not agree very wel with each other.

January 26, 1852.

RISE AND FALL OF WATER IN LAKE ERIE.

MESSRS. EDITORS: You published last spring th details of Colonel Stockton's daily register of lake levıls at this place for 1845–6, and some other statistics on the same subject. I then stated that all of the re-corded observations, at various points on the lake, were not in my possession, but at some future time, when they should be collected and compared, I would give you the *general result.*

So far as I have been able to discover, I now have them all; some taken at Detroit, others at Black Rock and Buffalo, and others here; but none of the daily observations extending through more than two years at either place. I propose to give them for each of those places, by themselves, in the terms in which they are kept, referring to the local zero or *plane* of reference. This will occupy about a column of your paper, and will be useful to subsequent observers at those points. From the whole, some *general conclusions* can be drawn, showing 1st, the mean *annual* rise and fall; 2d, the greatest *temporary* fluctuations known; 3d, the greatest mean and total *yearly* fluctuations known. It will be seen from these tables how imperfect these observations have been hitherto, with the exception of the years 1839 and 1840–41 at Detroit; the years 1838, 1845–46 and 1851 at Cleveland, and 1840–41, 1850–51 at Black Rock and Buffalo.

It has not been easy to reduce the whole to a common standard, or *plane* of *reference* for all parts of the lake, because its surface is very seldom *level.* The observations at Black Rock present greater discrepancies than any other, but should not be discarded; and I mention it because future detailed observations may show some errors in my tables, arising from this cause.

The Cleveland and Buffalo zeros have been compared by several months' consecutive readings, by Mr. Lothrop and myself in 1851; Mr. Lothrop's daily, and mine as often as circumstances would permit.

The month of July, 1851, was very quiet, and at Cleveland the water did not vary to exceed (2) *two* inches during the month. The average is (1) *one* foot 11½ inches below zero. The average of about sixty readings at Buffalo is 9 feet 47-100, for the depth of water in the canal; and if the *water line* between Cleveland and Buffalo was level, as it should be, if it ever is, during that month of calms, the *Cleveland zero* is (11) *eleven* feet 42-100 *above* the Buffalo zero, or bottom of the Erie enlarged canal.

In Detroit, taking the months of July, August, September, and October, 1838, as frequently measured during those months, at both places, the water at Cleveland was at an average of (1) *one* foot (1) *one* inch below zero; and at Detroit (4) *four* feet (6) *six* inches below the water table of the hydraulic tower; and therefore the Detroit zero is (if the waters were level and adding the descent of the river) (3) *three* feet (5) *five* inches *above* the Cleveland zero, and (14) *fourteen* feet (10) *ten* inches above the bottom of the Erie Canal.

By the use of these figures, if readings of the stage of the water are had, at one of the cities above named, say for a month consecutively, and not at the others, and it should be desirable to ascertain the level, it could be arrived at very nearly by calculation.

For instance: we have for the summer of 1840 and 1841, the monthly mean at Black Rock, and no register at Cleveland, but could make out a *theoretical* table for these months, which would be nearer the truth than *a few straggling measurements*. So for those years in the local tables following, which are *wanting;* by a good average or yearly mean, taken at another part of the lake, a close approximation could be made. And hereafter, when the subject shall be more fully investigated, by watching the *rain uage*, and knowing the general quantity of water hich falls in the lake region, the stage of the water ıay be safely *predicted* for the coming year.

It would not be correct to make one or two readıgs in a month, and call these the *average*, and use ıem to calculate levels elsewhere. It is in this resect that the tables are defective; but we must take ıem as they are, and make the most of them; hoping ıat for future years they may be more full. If lightouse keepers, were required to note the state of the ·ater daily, the object would be sooner accomplished.

Instead of traditions, we should have accurate ata, and practical benefits as well as scientific in- ›rmation would be cheaply obtained. The construc- .on of harbors, wharves, and warehouses could be lanned with intelligence, as to depth of water and levation above it, as experience should fix, the mount and limit of fluctuation in the lake surface.

WATER LEVELS.—DETROIT RIVER.—1819 to 1850.

Taken from the Register of A. E. Hathan, Esq., Civil Engineer.
Zero—the top of Water Table of Hydraulic Works.

Date.	Surface below zero.	Height above 1819	Mean difference from year to year.		Month of high and low water.		Greatest temporary fluctuations within the year.		Greatest difference.
			rise.	fall.	high	low.	high	low.	
1819, June, Lowest known stage	*8,45								
1828 and 1830, June	6,45	2,00	2,00		J'ne	Feb. Mar.		6,86	
1836 "	4,79	3,66	1,66						
1837 "	4,20	4,25	0,59						
1838, August 21 Highest known stage	3,20	5,25	1,00						
1838, average of five last months.	4,25	4,20		1,05	Aug	Dec. Jan.	3,20	6,86	3,66
1839, average of six measurements during the year	5,58	2,87		1,33	J'ly.	'39 Jan.	4,52	7,68	3,16
1840. 17 measurements	6,55	1,90		6,97	J'ly	Jan.	5,83	7,92	2,09

* Colonel H. Whiting.

Table Continued.

Date.	Surface below zero.	Height above 1819.	Mean difference from year to year.		Month of high and low water.		Greatest temporary fluctuations within the year.		Greatest difference.
			rise.	fall.	high	low.	high	low.	
1841.									
12 measurements........................	7,55	0,90		1,09	Aug	ja. fe Mar	6,61	7,63	1,02
1842.									
January 31..............................	7,58	0,87	mean of 8 years, 120				mean of 4 years.		2,96
June 1st..................................	7,86	0,59							
Lowest stage since 1819..............									
1843.									
July 1st..................................	7,53	0,92							
1847.									
October 24..............................	6,13	2,32							
1849.									
March 21................................	6,28	2,17							
May 3d....................................	5,53	2,92							
1850.									
July 5......................................	5,58	2,87							

Abstract of Water Levels at Cleveland, Ohio.
1796 to 1852.

The *measurements* commence at 1826. Prior to this date we have nothing but *estimates*, which were made by Mr. Alonzo Carter, Leonard Case, Esq., Mr. Levi Johnson, and others of the old settlers. A. Walworth, Esq., while government agent for the harbor works, noted the height of water occasionally, from 1826 to 1838, when by his request Mr. George C. Davies kept a daily register, from July to December of that year. The figure for January, 1837, is given by A. Merchant, Esq. From August, 1845, to August, 1846, we have Colonel Stockton's daily register, nd from that time, reading more or less frequent y myself, made as often as circumstances would per-nit.

Date and Remarks.	Below '38, or zero, ft. & in.	Mean yearly difference.		High and low month.		Greatest observed range within the year.		
		rise.	fall.	high.	low.	high.	low.	differ-ence.
1796	low.							
1798	"	slight.						
1802	5 ft.							
1806	low.		slight.					
1810–11	6 ft.		"					
1813		slight.						
1814		2 ft. 6						2 f. 6
1815, June } very	" 5 ft.		4 to 5		June.			
" Fall months } wet	" 2 ft.	3 ft. 0		Fall				
1816, Spring } years	" 2 ft.			June.				
819, late in summer very dry '17–'18	" 6 ft.		4 ft.		Aug.			
owest known stage								
822	" 5 ft.	1 ft. 0						
825	" 4 ft.	1 ft. 0						
826	2 ft. 10	1 ft. 2						
832	2 ft. 10	0 ft. 0				2 ft. 6	3 ft. 2	0 f. 8
833	3 ft. 2		0 ft. 4			2 ft. 8	3 ft. 8	1 f. 0
834, June	2 ft. 7							
835, "	2 ft. 10		0 ft. 3					
" September	3 ft. 4							0 f. 6
839, January	3 ft. 1							
" June	2 ft. 7			June.				
" September	3 ft. 6							0 11
838, May 10	1 ft. 0							
" June 12	0 ft. 4							
" June 25	0 ft. 00	2 ft. 7		June.				
ighest known stage								
838, July 12	0 ft. 3							
" Aug., daily register	0 f. 9½							
" Sept., " "	1 f. 2½							
" Oct., " "	1 f. 6½						1 ft. 11½	1 f. 11½
" December 1st	1 11½							
841, March 11	2 ft. 9							
842, April 14	2 ft. 5	....	....					
845, Mean of daily reg.						2 ft. 6	4 ft. 8	2 f. 2
" August	2 f. 8,3							
" September	21 f. 14							
" October	3 f. 16							
" November	3 f. 5,6							
" December	3 f. 9,5							
846, January	4 f. 0,6							
" February	4 f. 4,8							
" March	4 f. 5,8				March		4 f. 5,8	
owest since 1819								

Table Continued.

Date and Remarks.	Below '38, or zero, ft. & in.	Mean yearly difference.		High and low month.		Greatest observed range within the year.		
		rise.	fall.	high.	low.	high.	low.	difference.
1846, April	4 f. 1,2							
" May	3 f. 1,6					2 ft.		
" June	2 f. 10			June.		10 ft. 1		
" August	3 f. 0,7		0 f. 4,4	....				1 7
1848, October 10	2 f. 8							
1850, June 20	2 f. 11							
1851, January 6	3 f. 8				Jan.		4 ft.2	
" " 18	4 f. 2							
" June 5	1 f. 11	1 f. 0				1 f. 11		2 f.3
" " 15.	1 f. 11			June.				
" July—Mean of eight measurements in a l weather	1 11½			July. Aug.				
" August 3d	1 11½							
" " 7th	1 11							
" September 19th	2 f. 3							
" October—mean of 5 measurements	2 f. 5							
" November—12 mea..	2 f. 8							
" December—8 mea.	2 f. 12							

Mean annual difference of level (in four best ascertained years).......... 1 ft. 3 in.
Greatest annual difference of level........ 2 ft. 2 in.
Least annual difference of level........ 0 ft. 0 in.
Mean of observed fluctuations within the year (four best years)........... 2 ft. 0 in.
Greatest " " " .... 2 ft. 3 in.
Least " " " .... 1 ft. 11 in.
For 1845–6, the *mean* of the greatest *transient* fluctuation for 13 months, as shown by the daily record, arising from *winds* and *storms*, within each month is........ 1 ft. 5 in.
The greatest temporary *height* from local and sudden causes, such as storms, stood—below zero........ 1 ft. 10 in.
The greatest *depression* from like causes........ 5 ft. 00 in.

In certain ports of Lake Superior and Lake Michigan, such movements take place several times a day, sometimes within a few minutes of each other, the water rising and falling from half a foot to (3) *three* feet.

WATER LEVELS AT BLACK ROCK AND BUFFALO.—1819 to 1851.

'aken from the Buffalo Commercial Advertiser, and the Register of JOHN LOTHROP, *Civil Engineer—measurements reduced to the Buffa-lo* zero, *or bottom of the enlarged canal, which is* one (1) *foot below the mitre sill of the Guard lock at Black Rock.*

Date and Remarks.	Depth of water in canal.	Height above 1819-20.	Mean yearly difference.		Greatest observed fluctuations.		
	ft. & 100		rise.	fall.	low.	high.	dif. in the yr.
319-20—August							
owest known stage	6,30						
328—August.	8,80	2,50	2,50				
330 "	8,80	2,50					
336 "	9,80	3.50	1,00				
339 "	10,30	4,00	0,50				
338 "	11,40	5,10	1,10				
ighest known stage						Aug.	
339—							
May 11	9,50						
August	10,13	Mean.					
October	10,30	3,83		1,27		Oct.	0,80
340, Monthly average							
May	10,33						
June	10,90					June.	
July	10,33						
August	10,40	Mean of					
September	9,30	6 m'ths.					
October	9,10	2,60		1,23			1,23
341—							
May	9,50					May.	
June	9,24						
July	8,80						
August	8,90	Mean of					
September	7,75	6 m'ths.					
October	6,86	2,41		0,49	Oct.		2,64
owest since 1819							
342, May 5	9,90	3,60	0,40				
343 (estimate) May 15	8,96	2,66		0,94			
344, May 12	9,21	2,91	0,25				
345, " 15	9,30	3,00	0,09				
346, " 16	8,30	2,00		1,00			
347, " 16	8,80	2,50	0,50				
348, " 1st	8,46	2,16					
349, " 19	9,40	3,10	0,94	0,34			
350, " 8	8,96	2,66					
Iean of daily obs.)				0,44			
November	7,71				Nov.		
December	7,83						

Table Continued.

Date and Remarks.	Depth of water in canal.	Height above 1819–20.	Mean yearly difference.		Greatest observed fluctuations.		
	ft. & 100		rise.	fall.	low.	high.	dif. in the y.
1851—							
January	7,88						
February...............	7.86						
March..................	8,40						
April..................	8,47						
May....................	8,59			0,37			
June...................	9,34						
July...................	9,46					July.	1,75
August.................	9,34						
September..............	9,07	mean of					
October................	9,13	12 mths.					
November...............	9,21	2,21					

Mean annual difference of four best ascertained years................................ 0,89
Greatest annual difference... 1,27
Least annual difference... 0,37
Greatest temporary fluctuation within the year.. 2,64
Least " " " " 0,80
Mean of fluctuations.............. 1,40

The columns of "mean yearly difference" and of "greatest temporary fluctuations" do not in all cases represent those facts truly, on account of the diverse character of the registers, and the difficulty of combining all the facts in one table. For instance, the *annual* or mean yearly difference sometimes expresses the difference between the *same month* of two consecutive years, instead of the *mean* of the two years compared with each other. The measurements are so incomplete that this is often the best difference that can be had. In the column of "temporary fluctuations," or difference within the year, the figures do not show the extreme elevation or depression caused by *violent storms*, but a temporary high or low stage of a month or less.

The column of high and low *months* is given to

show the remarkable *uniformity* of high and low water, as to *season* of the year. At Buffalo and Black Rock there is less regularity than elsewhere, showing that winds, and the form of the coast, produce irregularities there that are local.

For (5) *five* years high water occurred there—in July, *once;* June, *once;* May, *once;* August, *once;* and October, *once.* For the winter months, when the water is low, the measurements are wanting.

At *Detroit* there are several years when there was little variation during two or three months. In *four* years (not consecutive, but those best ascertained) it was high in June, and July *three* times, and August *once;* and January, February, and March (which were about equal) *once.*

At *Cleveland*, in five years well ascertained, high water has been in June, *four* times; and June, July, and August (about equal) *once.* Low in January, *once;* December, *twice;* March, *once*, and August *once.*

So that out of fourteen cases where the highest months have been registered, it has been high in June and July (10) *ten* times, and in (10) *ten* cases has been *low* in December and January (6) *six* times. It is like other ponds and rivers, raised by the spring freshets, and lowered by frost and drought of winter.

I am unable to find in these tables any confirmation of the popular belief, that there is a "seven years' rise and fall" of the water in the lakes.

The Detroit registers show a continual rise, from 1819 to 1838, or *nineteen* years.

From 1838 to 1841, a continual decline of *three* years. In 1842, a slight rise. From 1842 to 1851, *eight* (8) years, a regular decline.

At the other extremity of the lake accounts differ as to the fact of the lowest water being in 1819 or

1820; but the difference between those years was slight. It was at the greatest height in June, 1838, at Buffalo, and in August, 1838, at Black Rock; so that for 18 or 19 years there was a steady rise. From 1838 to 1841 a continual decline of *three* years, the same as at Detroit. From 1841 to 1851, ten years, it has been fluctuating up and down, about as much rise as fall; that is, the total rise 2,75, total fall 2,56. At Cleveland, from 1796 to 1810–11, the water was low *fourteen* years. From 1811 to 1816, *five* years, rising very rapidly. From 1816 to 1819, *three* years, falling rapidly to the lowest known point. From 1819 to 1838, *nineteen* years, a steady *rise*. Since 1838 there has been a steady decline to March, 1846, the lowest since 1819; and since 1846 considerable fluctuation and irregularity. Among all these periods there is *none* of *seven*, and just *one* of *fourteen* years.

These tables demonstrate the effect of the change of *seasons within the year* upon the stage of water. They show an *annual* rise and fall entirely independent of the general height of water fluctuating a certain distance within each year, whether the general surface is high or low.

The average or mean annual difference of four well ascertained years at Cleveland, is	1 ft. 3 in.
Of (8) years at Detroit	1 ft. 2½ in.
Of (4) four at Buffalo	0 ft. 10½ in.
Mean of all annual differences	1 ft. 1½ in.

No observer of Registers has discovered a *daily or lunar tide*.

The greatest range of *general surface* of the lake at Detroit was between August, 1838, and June, 1819	5 ft. 3 in.
At Black Rock, between August, 1838, and August, 1819 or 1820	5 ft. 3 in.
At Buffalo, between June, 1838, and August, 1819	5 ft. 3 in.
At Cleveland, between June, 1838, and the fall of 1819	5 to 6 ft.
The greatest *extreme* and *temporary* range noticed in the Register occurred between the same years, are given as follows:	
At Detroit	6 ft 8 in.
At Black Rock	7 ft. 1 in.
At Cleveland ......about	7 ft. 0 in.

At this time Lake Superior is unusually high. During the summer of 1851 it was about (3) *three* feet above the general level of 1847, when it was unusually low. Lake Huron is stated to have been ($2\frac{1}{2}$) *two and one-half* feet higher at the Detour light than in 1847. Lake Michigan is reported to be higher the past autumn than for several years. The upper lakes being full must discharge a large surplus into Lake Erie and Ontario, which will be felt in the coming spring. If we have a cold and wet season, it will operate at that time to assist the flood of water from the north, in raising the surface of our lake. If it is warm and dry, it will counteract the effect of a large supply through the Detroit River. Instead of regarding the rise and fall of water in the lakes as a mystery, it is rather to be wondered that there is so *little* fluctuation. Their stability is dependent entirely upon the regularity of the seasons within the lake country, and if there should be a combination of wet and cold years, wherein the fall of rain should be great, and the evaporation small, there might be a rise or fall exceeding anything we have on record.

COAL AND IRON TRADE OF THE OHIO VALLEY.

[Merchant's Magazine, May, 1847.]

THE coal and the iron fields that exist between the Alleghany Mountains and the Mississippi River are commensurate in extent, because the strata of ironstone and coal alternate with each other. Iron is, it is true, a mineral not confined to one rock or formation, but ranges from the primitive rocks, up through the sedimentary strata, to the recent alluvion. But the world over, it is a geological law, that the coal-bearing rocks are composed in sensible quantities of the ores of iron; so that an explorer, having discovered that he is in the midst of the carboniferous system, expects to find beds of iron with as much confidence as he expects coal.

This metal may not be so abundant in all parts as to be of economical value; but strata of greater or less thickness may be relied upon as forming part of the regular geological structure of the country. Thus we may foresee the immense product of iron that the western coal fields will, of certainty, yield to posterity.

During the past two years, four furnaces have been built on the Mahoning Canal that use raw bituminous coal, in lieu of charcoal, in reducing ores. Three of them are in the county of Mahoning, Ohio, at Youngstown and Lowell, and another at Tallmadge, near Akron, in Summit county. Two of them have been in operation long enough to test the project, and the results are, that pig metal can be produced in this way at less cost than with charcoal. The consequences of this experiment, and its success, are prodigious. Ores that are called "harsh" by the founders, containing silicious matter, and therefore refractory and expensive, are found to be more easily reduced by the concentrated heat and blast of the coal furnace than by the charcoal stack. The limit to the manufacture of iron is thus not restrained by the want of timber; nor are the woodlands of the country destroyed to supply the furnaces. Mineral coal being literally inexhaustible, the only bounds to the production of iron are the supply of ore and the demand for the article. Coal and coal lands become thus of higher importance in the economy of a country, and of more local value.

Geological investigations have gone so far as to determine, with general accuracy, the boundaries of the Alleghany coal field. It is of an oblong form and somewhat irregular, the longest axis extending north-east and south-west, from the neighborhood of Meadville, Pennsylvania, to that of Huntsville, Alabama, nearly 600 miles in length. It is widest at the northern part, tapering to a point at the southern extremity. Its breadth is greatest at Pittsburgh and Wheeling, where the Ohio River occupies a central position, and its thickness at the centre is estimated at 2,000 to 3,000 feet. By this is meant, that all the strata of sandstone, shale, coal, limestone, and ironstone, that compose the coal series or "formation," from the conglomerate, the base of the formation, to the top of the same, are, inclusive, so many feet thick.

The region occupied by these strata is called a basin, or a coal basin, because the strata plunge towards a common centre, or central line; so that a boring, or well, made in the valley of the Ohio River, at or near Wheeling, would pass through 2,000 or 3,000 feet of these rocks before reaching the conglomerate, which is seen at the surface, at Akron on the west, and at the summit of the Alleghanies on the east. In *physical level*, the eastern outcrop of the lowest bed of coal is *higher* than the surface of the upper beds of coal; but in geological order of superposition, it is *lowest* of all. For instance, the bed which is worked near the station-house of the Portage Railroad is the one at the bottom of the series, but is 2,000 feet above tide-water. The beds in the neighborhood of Wheeling are higher up in the series, and

2,000 or 3,000 feet above the continuation of the Portage summit bed, extending westward to that place; but the Ohio River is here only about 640 feet above the ocean, and the hills adjacent about 500 feet more. The bottom of the coal strata is therefore 1,500 or 2,000 feet below the surface of the ocean. Such is the result of a gradual plunge, continued through long distances; the lower bed of coal, having descended from the summit of the Alleghany Mountains, 2,000 feet above the sea, to a point as many feet below it, and then rising towards the west, appears at the surface, on the other side of the basin, at Akron and Newcastle, 900 to 950 feet above the ocean level. The distance between the two sides of the field or basin, on its lesser axis, is about 200 miles.

The entire number of coal and iron strata embraced in this mass is not known; but if we could penetrate it from top to bottom, or make a vertical section, as we are enabled to do by observing the face of the rocks at various points, we should probably find at least fifty strata of coal, and more than twice that number of ironstone, lying in regular order one above another. Of these, twenty or twenty-five of the coal strata might be workable; or say, three feet thick and upwards to six feet; and of the iron, more than one-half would pay for stripping at the edges around the hills. In Lawrence county, Ohio, on the western verge of the field, where the strata dip gently to the eastward, in the vertical space of about 800 feet, there are seen *four* workable strata of coal, and *eight* of iron, with many more regularly stratified beds of less thickness. Here, a bed of coal less than three feet is not considered valuable; and ore is thought worth stripping when an inch may be had by removing a *foot* of earth. All parts of this great eld may not be as rich, but some are known to be ıore so; and iron is found, in several instances, out-:de of the coal region. Here is an area, therefore, ırger than all England and Scotland, over which fur-aces may be supported, if a demand for iron could y possibility arise equal to such a capacity for pro-uction.

On the Lower Ohio, in Kentucky, Indiana, and linois, is another basin, or field, of coal and iron, of ırge dimensions, but detached from the one above ɔticed. It is also oval in form, and more regular ıan the Alleghany field; its greatest length being in north-westerly and south-easterly direction, from ıe north-west angle of Illinois, passing the mouth of ıe Cumberland to the south line of Kentucky, say)0 miles. It embraces a large portion of Illinois, veral of the south-western counties of Indiana, and ur or five of the Green River, Tennessee, and Cum-ırland River counties, in Kentucky. But because a rge part of the tract is level, the strata do not crop ıt advantageously for mining; and their edges are en principally on the banks of streams and collat-al valleys that put out from the main ones. The ineral power of this region is but little understood. t Hawesville, and a few other points, coal is fur-shed for steamboats and taken to New Orleans.

Beyond the Mississippi, in Missouri and Iowa, and en to the sources of the Arkansas, coal is known to :ist; but as yet it is not explored so as to define its nits or value, or to determine whether it is a part the Illinois field, or of one or more separate basins. In Michigan, also, there is a basin, including about ıe-half the lower peninsula; but the strata are thin, ıd the position retired from navigation. And in ad-tion to the iron ore, necessarily attendant upon such ımerous and extensive beds of coal, there are, extending from Lake Superior, with occasional intervals, through Wisconsin, Missouri, Arkansas, and Texas, masses of iron, in the primitive and volcanic beds, that exist along a line from Michigan to Mexico.

By the census of 1840, there were, in the United States, 804 furnaces, producing annually 285,903 tons of pig metal and castings. There were also 795 forges or refineries, turning out 197,233 tons of malleable iron.

The *bituminous* coal raised was 27,603,191 bushels; which, at 70 lbs. to the bushel, is 966,111 tons; of anthracite coal, 863,489 tons.

On account of the increased demand, and also in consequence of the introduction of the hot blast, by which the yield of a furnace is increased from one-third to one-half, without knowing the number of the furnaces and iron-mills erected since 1839, I think it safe to allow 25 per cent., or one-quarter, for the enlarged production of 1846 over 1839: —

	TONS.
That is, for pig metal in the United States,	358,024
For malleable iron and iron rails,	246,531

The increase in the quantity of bituminous coal, raised and consumed, is still greater—probably 50 per cent., or one-half.

In February, 1846, the descending coal trade of the Ohio was estimated at 12,000,000 of bushels, or 480,000 tons.

In 1840, there were received at Cleveland, by the Ohio Canal, 6,032 tons; in 1846, 31,283 tons.

This is not all the coal consumed upon the lakes; for the Erie extension, now in operation, delivers at Erie, in Pennsylvania, a large amount, probably 12,000 tons. At the time of the census of 1840, the mines on the Lower Ohio had scarcely been opened;

and the steamboats on the Ohio River, like those on the lakes, had not become habituated to the use of coal.

With all these indications of increased consumption at the west, where the principal beds of bituminous coal exist—for the United States, I think it safe to put the augmented business in that time at 50 per cent. The new use in stack furnaces, and the increased use in rolling-mills and forges, add much to the already monstrous application of this fuel. We will therefore state the present amount of bituminous coal raised, which is principally at the west, at 1,449,161 tons.

This does not probably show more than one-third of the consumption of the United States, including the anthracite and imported coals. At that rate, the total consumed in the United States would be 4,347,748 tons, or about the same as that of France, in 1841.

It may appear singular, but it is nevertheless true, that in the experiments upon the heating power of coal, made at Washington, in 1843–44, at the expense of the government, under Professor Johnston, only *three* specimens were taken from the west of the Alleghany Mountains, out of *fifty-eight* specimens operated upon. Of the three, one was from Pittsburgh; one from Cannelton, Indiana; and one from the New Orleans coal-yards,—its origin not known. We are therefore still without the benefit of most of the splendid results that flow from these experiments.

The practical value of the coal, everything else being equal, is its capacity to make steam; and the rule of the experimenter was to determine the quantity necessary to convert *one cubic foot of water* into steam:—

	LBS
The Pennsylvania and Maryland free burning coals required for that purpose	7.33

	LBS.
The anthracite	7.71
Richmond	8.20
English and Western	8.97

In regard to western coals, the number of specimens was too small to give much value to the conclusion, in regard to their heating power. It is satisfactorily settled, however, that the heating power is not in *direct proportion* to the carbon of the coal; for although the anthracite is nearly pure carbon, it stands below the free burning Maryland and Pennsylvania coals that contain bitumen.

According to Professor Silliman, the George's Creek coal, Maryland, of which four specimens were analyzed, contained 18⅓ per cent. bitumen; and it is this and the kindred kinds which, according to Professor Johnston, stand at the head of the list. It is well known that, in Pennsylvania, there is a regular gradation from anthracite to bituminous coal, as we proceed from Mauch Chunk towards Pittsburgh. The Ohio coals contain, in general, a larger amount of bitumen than those of the eastern edge of the field on the summit of the Alleghanies; that is to say, from 30 to 40 per cent.

Reducing the bitumen to its elements, the Ohio coals, as far as analyzed, give about 81 per cent. carbon, while the English coals have about 73 per cent. Professor Johnston ranges the English and western, according to their heating effect, about the same.

Let us now refer to the return of coal and iron for the whole United States, by the census, and compare the proportion of both due to the Ohio Valley, by which I mean the region drained by its waters.

	No. of furnaces, and tons of cast iron.		No. of forges, and tons of iron.		Bush. of bituminous coal.
7estern District of Pennsylvania,	134	53,101	67	63,431	11,620,654
" " Virginia........	30	10,892	38	3,721	8,073,364
ennessee.	34	16,128	99	9,673	13,942
.entucky.	17	29,206	13	3,637	1,158,167
hio.	72	35,236	19	7,466	3,597,769
ndiana.	7	810	1	20	242,000
linois.	4	158	0	......	461,807
or the Ohio Valley.	298	145,531	337	87,948	25,167,703
or the United States.	804	286,903	795	197,233	27,603,191

By these footings, about *one-half* the iron made n this nation is turned out upon the waters of the)hio, and almost the whole of the bituminous coal.

I have no means of stating the quantity of anthraite coal now raised, or of giving the probable increase ince 1839–40. In Ohio, since the above enumeraion was taken, there have been at least *eight* furnaces rected, and in Kentucky *four;* most of them hot last furnaces. There has also been an increase in vestern Pennsylvania. Throughout the west, geneally, it may be asserted, that the number of works nd the product of individual works have increased in greater proportion than east of the mountains. If his is true, the *relative* product of the Ohio Valley nd of the nation, at this time, would be different rom that shown in the preceding table, and the diference would be in favor of the west.

The duty on coal, under the act of 1842, was $1 75 per ton. From September 1st, 1845, to March st, 1846 (six months), New Orleans received by the iver 300,000 bushels, which it was supposed might e met by imported coal, under a duty of $1 per ton. The act of 1846 fixes upon coal a duty of 30 per ent. ad valorem.

It is an article that varies greatly in price, at diferent places, and almost as much at the same place t different times.

	CENTS.
At New Orleans, by retail, per bushel, from	12 to 18
Cincinnati	9 to 15
Wheeling	3 to 5
Pittsburgh	4 to 5½
Cleveland	8 to 12
Philadelphia (Feb., 1846), bituminous,	20 to 22
New York, Nova Scotia coal . . .	18 to 21½
New York, English coal . . .	23 to 25

These prices are, of course, mere approximations.

By the experiments of Professor Johnston, the effect of anthracite, in generating steam, is not greatly superior to that of bituminous coal; and consequently, for household consumption, the bituminous, if furnished at about the same price, will work its way into favor. The cheerful brightness of its flame is, to many persons, more than a compensation for the difference in heat.

At Albany, there is already a small demand for coal from Lake Erie, at anthracite prices—say $6 to $7 per ton. It is more than probable that, after the Erie Canal is enlarged, this article, like the wheat, flour, and pork of the lakes, will become an important item in western trade.

At Cleveland and Erie it can be delivered in bulk on large contracts, at $2 25 and $2 50 per ton of 2,000 pounds.

Half a ton, or fourteen and a quarter bushels, of bituminous coal, is more than equal to a cord of four-foot wood; in fact, some regard ten bushels, and others twelve bushels, as equal to a cord.

There is therefore seldom, if ever, a time, even in the greatest scarcity of coal in market, when coal is not *cheaper* than wood as a fuel; ordinarily, it is about *one-half* less. This fact, taken in connection vith its greater safety, less trouble, uniformity of emperature, and the increasing scarcity of timer, explains why mineral fuel conquers every other, verywhere, and works its way into all departments f life.

In the coal regions, for most purposes of power pplied to machinery, it is crowding hard upon the ld method of water-wheels, substituting the steamngine in their place. For such uses, the bituminous oal seems to please best, on account of the readiness vith which it may be set on fire, and thus a quick team is obtained. It will undoubtedly always bear a igher price in the principal eastern cities than anhracite.

The interior of the great Alleghany coal field nay be thought too remote from the principal communicatians to be of anything more than a local alue. But in Virginia and Kentucky, the Cumbernd, Kentucky, Licking, and Kenawha Rivers, exend far into the coal measures, and in high water ne arks or flat boats are enabled to descend with a ull load. The Monongahela and the Youghogheny kewise cut through coal strata their entire length, nd the same may be said of the Muskingum.

From the sources of the branches of the Upper hio, to the neighborhood of Portsmouth, all the treams flow over beds of coal, or have worn their way hrough them in the course of ages.

The "Erie extension" is cut in the coal strata, nd also the Mahoning, and the Sandy and Beaver anals. The Ohio Canal, from Akron to Dresden, is n the same series; and thence to Portsmouth skirts s western edges. The Hocking Canal is also in the oal region.

At present, the principal mines on the river are t Hawesville, Pomeroy, Wheeling, Pittsburgh, and thence to Brownsville. It is from these points tha the flat boats are filled; but at a hundred other place can coal be taken, in any quantity, with equal facility as soon as it shall be needed.

COLONEL BOQUET'S EXPEDITION.

Extracts from a discourse delivered before the Young Men's Literar Association of Cleveland, December 17, 1846, *on the subject of Coi onel Boquet's Expedition into the Muskingum country, in October*, 1764

[Cleveland Herald, December, 1846.]

To understand the situation of the Indian tribes against whom the expedition of 1764 was sent, w must remember that it was before the Revolution, an that great changes had then recently taken place o this continent. It was only a little more than a yea since the English had been put in possession of Cana da and the left bank of the Mississippi. The ol French war, as it is called, had but just closed, i which the English were victorious, and by the treat of Paris, in February, 1763, England received all th country of the lakes and the St. Lawrence.

There had been a long dispute between these na tions about the regions south of the lakes and wes of the Alleghany Mountains, which resulted in th French giving up everything in North America as lo as the river Iberville, which was a mouth or bayou o the Mississippi above New Orleans.

When this war broke out in 1752, the French ha

ılitary possession of a line of posts from the Bay of unday across the Connecticut River to Crown Point, swego, Niagara, Presque Isle, now Erie, and Fort itt, then called Fort Duquesne. They had succeeded ı gaining the confidence of the Indians, who fought ı great numbers against the English and with the rench.

By the treaty the garrisons of all the places I ave just named, and also those of Detroit, Mackinac, t. Joseph's, Vincennes, and about thirty more, march-l out, and the English troops entered.

The Indians were very much displeased when they ıw the English taking possession of their country, for ıey preferred the Frenchmen, who had been their iends and traders more than one hundred years, and ıd married Indian women.

A noted chief of the Ottawa tribe, known by the ıme of Pontiac, formed the resolution to destroy all ıe English frontier posts at one assault, in which he as encouraged by the French traders. This Indian arrior, whose name ranks with those of Philip, Red ıcket, Little Turtle, and Tecumseh, seemed to pene-ate the future, and foresaw the dangers that threaten-l his people by the presence of the whites. He was ossessed of an active mind, of physical energies not ommon even among Indians, and a stirring eloquence hich won him authority and influence. He resided ; the west end of Lake Erie, and from this central osition he visited the western tribes.

He represented to them that they were now strong ıd the English were weak, but that soldiers were con-nually arriving, so that in a few moons the Indians ould be weak and the white men strong; that the esigns of the whites were now manifest; one nation ıd without their consent transferred them to another, ıd no one knew how long it might be before they would be again sold like slaves to some other powei that they were either to be annihilated or enslaved, ar that it was better to die fighting than to suffer eithe

He succeeded in forming an alliance with tl Ottawas, having 900 warriors; the Pottowotomie with 350; Miamies of the Lake, 350; Chippewa 5,000; Wyandots, 300; Delawares, 600; Shawanes 500; Kickapoos, 300; Ouatanons of the Wabas 400, and the Pinankeshaws, 250; in all, able to mu ter 8,950 warriors.

This may be called the "First Great North-wes ern Confederacy" against the whites. The *secon* took place under Brandt, or Thayandanegea, durii the Revolution, and was continued by Little Turtl the third under *Tecumseh*, in the last war.

Pontiac's projects were brought to a focus in tl fall of 1763, and the result was nearly equal to t design. The Indians collected at all the north-wes ern forts, under the pretence of trade and friendly i tercourse, and having killed all the English trade who were scattered through their villages, they ma a simultaneous attack upon the forts.

The forts at La Bay (probably Green Bay) a at St. Joseph's, on Lake Michigan, were taken, a also Fort Mackinaw. Fort Miami on the Maum River, and that of the Ouatanons on the Wabas Fort Junendot on Sandusky Bay, Fort Presque I (now Erie), and two subordinate works connecti Erie and Pittsburgh were surprised, and the most h rible fate inflicted upon all Englishmen found witl them. Detroit, Fort Pitt, and Niagara were able resist the assault; but Pontiac having much of l force released by success at other posts, concentrat it upon Fort Pitt and Detroit, as the most obnoxio and important posts. Both these places were inve ed by an Indian blockade and siege.

The inhabitants of Pennsylvania and Virginia were now subject to great alarm, and frequently robberies and murders were committed upon them by the Indians, and prisoners were captured.

General Gage was at this time the Commander-in-Chief of the British forces in America, and his headquarters were at Boston. He ordered an expedition of 3,000 men for the relief of Detroit, to move early in the year 1764. It was directed to assemble at Fort Niagara, and proceeded up Lake Erie in boats, commanded by General Bradstreet.

The other was the expedition I design principally to notice at this time. It was at first composed of the 42d and 77th regiments, who had been at the siege of Havana, in Cuba, under the command of Colonel Henry Boquet. This force left Philadelphia for the relief of Fort Pitt, in July 1763, and after defeating the Indians at Bushy Run, in August, drove them across the Ohio. It wintered at Fort Pitt, where some of the houses built by Colonel Boquet may still be seen, his name cut in stone upon the wall.

General Gage directed Colonel Boquet to organize a corps of 1,500 men, and to enter the country of the Delawares and the Shawanese, at the same time that General Bradstreet was engaged in chastising the Wyandots and Ottawas of Lake Erie, who were still investing Detroit. As a part of Colonel Boquet's force was composed of militia from Pennsylvania and Virginia, it was slow to assemble. On the 5th of August, the Pennsylvania quota rendezvoused at Carlisle, where 300 of them deserted. The Virginia quota arrived at Fort Pitt on the 17th of September, and uniting with the provincial militia, a part of the 42d and 60th regiments, the army moved from Fort Pitt on the 3d of October.

General Bradstreet, having dispersed the Indian

forces besieging Detroit, passed into the Wyandot country by way of Sandusky Bay. He ascended the bay and river as far as it was navigable for boats, and there made a camp. A treaty of peace and friendship was signed by the chiefs and head men, who delivered but very few of their prisoners,

When Colonel Boquet was at Fort Loudon, in Pennsylvania, between Carlisle and Fort Pitt, urging forward the militia levies, he received a despatch from General Bradstreet, notifying him of the peace effected at Sandusky.

But the Ohio Indians, particularly the Shawanese of the Scioto River, and the Delawares of the Muskingum, still continued their robberies and murders along the frontier of Pennsylvania; and so Colonel Boquet determined to proceed with his division, notwithstanding the peace of General Bradstreet, which did not include the Shawanese and Delawares.

In the march from Philadelphia to Fort Pitt, Colonel Boquet had shown himself to be a man of decision, courage, and military genius. In the engagement at Bushy Run, he displayed that caution in preparing for emergencies, that high personal influence over his troops, and a facility of changing his plans as circumstances changed during the battle, which mark the good commander and the cool-headed officer. He had been with Forbes and Washington when Fort Pitt was taken from the French.

The Indians who were assembled at Fort Pitt left the siege of that place, and advanced to meet the force of Boquet, intending to execute a surprise and destroy the whole command. These savages remembered how easily they had entrapped General Braddock a few years before by the same movement, and had no doubt of success against Boquet. But he moved always in a hollow square, with his provision train and his cattle in the centre, impressing his men with the idea that a fire might open upon them at any moment. When the important hour arrived, and they were saluted with the discharges of a thousand rifles, accompanied by the terrific yells of so many savage warriors arrayed in the livery of demons, the English and Provincial troops behaved like veterans, whom nothing could shake. They achieved a complete victory, and drove the allied Indian force beyond the Ohio.

The principal authority relating to this expedition, and also the one to the Muskingum River, is a work by Captain Thomas Hutchins, of the 60th regiment of foot. In 1765 he published a small quarto, in London, with a map of the country now forming the State of Ohio. Captain Hutchins accompanied both expeditions as a military engineer, and has left a handsomely written and extremely important book, and one which is now very rare.* The only copy I have seen was at the Louisville Mercantile Library. Captain Hutchins wrote "A description of the Ohio, Scioto, Kenhawa, Wabash, and Illinois" Rivers, which was published in London, November, 1778; and also an "Account of Florida, Louisiana, the Mississippi, &c.," published at Philadelphia, 1784. These are attached to Captain Gilbert Imlay's America, London, 1797.

Hutchins, in the first of the above works, says, "The *Cayahoga* is muddy and not very swift, but obstructed with falls and rifts. Here are fine uplands, extensive meadows, oak and mulberry fit for ship building; walnut, chestnut, and poplar for domestic uses. Cuyahoga furnishes the best portage between

* Colonel Peter Force, of Washington, has discovered that this work was composed by the Rev. William Smith, of Philadelphia, from the notes of Boquet, Hutchins, and others.

the Ohio and Lake Erie, and at its mouth it is deep and wide enough to admit large sloops from the lake. It will hereafter be a place of great importance."

To understand the full merit of the men and officers employed upon military duties in this country, at that time, we should remember that no roads, settlements, or other communications existed along the route. Occasionally there were found Indian paths or trails leading from the Ohio River to the lakes; but for an army to pass with its baggage wagons, and its artillery, ample highways were to be cut and bridges constructed. These troops were directed to make an incursion, not through a country of farms and villages, but to penetrate a vast forest, tenanted all over with a fierce and excited savage enemy, one who knew no rule of warfare but that of murder; or if prisoners were taken, they were reserved for a slow death, to which all imaginable tortures were added.

When Boquet's command set forth upon their march, they stood on the same ground where the unfortunate victims of Braddock's defeat were burnt alive, with every circumstance of suffering that their heathen captors could invent. They knew full well the fate that awaited them. They advanced cautiously, surveying a road, and measuring each day's travel with a chain. The detachment had proceeded but a few miles along the north bank of the Ohio, when they found themselves closely watched by the spies of the Delaware tribe. The old stratagem of Indian warfare, of forming an ambuscade, and of distracting the command by drawing off portions in pursuit, although attempted, did not succeed. The English troops were always in close order and ready for battle. They passed through *Logstown*, a village of the Shawanese and Delawares, where the French had a trading post, but the place was now abandoned. At the mouth of

Big Beaver, now called Mahoning River, the troops crossed by a ford. Here there was a trading post of the French, and an Indian town situated near where the town of Beaver now stands. But the place was deserted, and the remains of horses which the Delawares had killed were there.

On the 6th of October the troops reached Beaver Creek, which we call little Beaver, and the next day penetrated the wilderness seven miles and one hundred and thirty-seven rods. Their route passed up the east branch of the Little Beaver and across the highlands to the waters of Yellow Creek through an open bushy country. Here the Indian trail forked, one branch bearing southward to the lower towns, the other continuing westward towards the Tuscaroras and Sandusky villages.

On Wednesday, the 10th, the army after a march of seven miles and sixty rods, cross the Yellow Creek, and the next day reach the waters of Sandy Creek, which was called Big Sandy, a branch of the Muskingum. Passing down the north bank of the Sandy, they cross the Nimishillim, a creek on which the village of Canton, in Stark county, is now built, but called by Hutchins the *Nemenshehelas*. On Saturday they reach the Muskingum, at a village of the Tuscaroras, having a strength of 150 warriors, but abandoned, and the huts in ruins. They cross at a ford, and encamp on a beautiful plain where the village of Bolivar now stands. On Sunday they remain in camp.

When Colonel Boquet set out from Fort Pitt, he sent two messengers through the country with a despatch for General Bradstreet at Lake Junendot. During the day, while the army were taking rest after the fatigues of the march, these messengers came into camp. They stated that the Delawares, at a town about sixteen miles distant, had made them prisoners and had taken them to their town. But when the In dians saw how large an army was advancing, and witl what regularity and caution it moved forward int(their country, they determined to send the messenger: back, and directed them to ask for peace.

It should be remembered that no movement wa: made on the part of Colonel Boquet with which th(Indians were not fully acquainted. They knew th(exact strength of his party, and kept an eye upon hin from the adjacent hills and thickets, every day an(every hour of the march. They were also well ac quainted with the man sent to chastise them, and knev him to be determined on no half-way measures. H(had so far escaped from their hands, and they sav him seat himself on the banks of the Muskingum an(commence a fort. By Tuesday, the 16th of October the commander had erected a stockade two miles an(forty rods below the ford, at a ravine, and had com pleted his arrangements against a surprise. On thi: day six chiefs appeared in camp, and stated that th(head men of their tribe were assembled about eigh miles distant, and wished to come in and make a treaty The Colonel and a large portion of the troops proceed ed to a bower, and the chiefs came in, sat down upoı the ground and began to smoke. There were fifteeı Seneca warriors, with their chief, *Kigashuta.* Of th Delawares there were twenty warriors and two chiefs *Custaloga*, of the *Loups*, or the *Wolf* tribe, and *Beav er*, of the *Turkey* tribe. On the part of the Shawa nese were six warriors, and *Keissinautcha*, their chiel Kigashuta, Little-Heart, Custaloga, and Beaver wer the orators. They excused themselves for their lat outrages in the usual manner, throwing all the blam upon their young men and upon other tribes. Eight een prisoners were delivered to Colonel Boquet, an(

suing for peace in the most abject terms, they promised to deliver all the remaining prisoners, eighty-three in number. An answer to their speeches was set for the next day; but as it was then rainy the meeting was not held until the second day after, when the Colonel made a bold and decided reply.

He recapitulated their late murders, and particularly the destruction of four of his messengers, whom they had killed. He accused them of violating all their former engagements, saying—"You have promised at every former treaty, as you do now, that you would deliver up all your prisoners, and have received every time on that account considerable presents, but have never complied with that or with any other engagement. I am now to tell you, therefore, that we will no longer be imposed upon by your promises. This army shall not leave your country till you have complied with every condition that is to precede my treaty with you. If I find you faithfully execute the following preliminary conditions, I will not treat you with the severity you deserve. I will give you twelve days from this to deliver into my hands, at *Wakatomaka*, all the prisoners in your possession, without any exception, Englishmen, Frenchmen, women and children, whether adopted into your tribes, married, or living amongst you, under any denomination or pretencé whatsoever, with all negroes. And you are to furnish the said persons with clothing, provisions, and horses to carry them to Fort Pitt."

To this the Shawanese in a dogged manner replied that they would submit to whatever the other tribes should agree. The chief of the Senecas, addressing the nations, desired them to comply with their engagements, in order "that they might wipe away the reproach of their former breach of faith, and convince their English brothers that they could speak the truth."

On Monday the 22d the troops broke up camp, and proceeded down the west bank of the Muskingum toward the Wakatomaka towns about the mouth of the Whitewoman. The deputations accompanied them as guides.

They reach the highland one mile north of the mouth of the Walhonding or Whitewoman, on Thursday, and make a camp. The distance of this point from the mouth of Big Beaver or Mahoning river, by the route of the army, is one hundred and one miles and eighty-three rods.

Colonel Boquet caused a stockade to be built, with four redoubts, and erected cabins and storehouses, determined to wait for the arrival of the prisoners.

After two days, Custaloga sent word that they were coming, and so did the Shawanese. The latter tribe had a terrible notoriety among the whites, not only for their ferocity in war, but for treachery and cruelty to prisoners. They had been seated upon the Susquehannah, and driven thence by the Six Nations, had gone to North Carolina, and from thence returned to the Ohio River near Shawneetown, Illinois. The other tribes would not allow them a fixed residence among them, on account of their piratical habits; but gave them permission to remain for a time or the Scioto, and afterwards on the great Miami.

Captain Hutchins lays down twenty-four Indiar towns on his map, which were situated on the water of the Mahoning, the Cuyahoga, Muskingum, and Scioto. The troops had now reached a central positioi among these towns, and consequently their influence over the tribe was at its height.

During the previous winter Colonel Boquet had sen a letter to the French Commandant, at Fort Char tres on the Mississippi. The Shawanese had detaine this message, and he now sent one of their number t get it. On Sunday the messenger returned without the letter, but said the prisoners were coming in. Here the commandant received a despatch from General Bradstreet, written and sent from the Sandusky River, which was brought in by a Caughnewaga chief named Peter, with twenty warriors. It stated that Bradstreet had made a general peace with the western Indians as far as the Illinois, but that the Delawares and Shawanese were not included. Peter said that very few prisoners had been delivered at Sandusky, and that some of them had been killed. Prisoners were coming every day to the fort at Walhonding. Custaloga and Beaver said that they had produced all of theirs but twelve, promised to deliver those, and desired to shake hands. But Colonel Boquet refused to speak to them till all were brought in.

On the 9th of November 206 prisoners, including women and children, had been delivered, of whom 32 men and 58 women and children were from Virginia, and 49 males and 67 females from Pennsylvania.

Signs of the approach of winter began to appear, the trees shed their leaves, and the raw winds of the north moaned among their branches. The troops began to look with anxiety for the order to return, and the commander learning that the remaining prisoners, about 100 in number, were absent on a hunting expedition, saw they could not be expected that season. He determined to close the affair, and make his winter quarters at Fort Pitt.

A conference was held on the same day in a rude shed made of logs and branches. Captain Hutchins appears to have been acquainted with our illustrious countryman, Benjamin West, for Mr. West made a design of this conference, which was engraved and attached to Hutchins' work. Like Mr. West, Captain Hutchins in the Revolution joined the whigs, and afterwards became the geographer of the United States, under the Confederation, and appears to have been the *first man* who devised the plan of surveying the public land in sections of one mile square. At this conference the warriors delivered the last of their prisoners and asked for peace. Colonel Boquet replied that he was not empowered by his king to make peace, but only to make war; that Sir William Johnson was intrusted with the power to give peace, and in the mean time "you will deliver me two hostages for the Senecas and two for the Custalogas tribe, to remain at Fort Pitt as security that you will commit no farther hostilities or violence against his Majesty's subjects, and when peace is concluded these hostages shall be delivered safe back to you. Secondly, the deputies you send Sir William Johnson must be fully empowered to treat for you and your tribes, and you shall engage to abide by whatsoever they shall stipulate."

It was agreed that Captain John and Captain Pipe, two hostages then at Fort Pitt, should be immediately returned to them. After this, Colonel Boquet shook hands with them for the first time, which gave them great gratification.

On the next day, in the presence of the Senecas, the Turkey band of the Delawares, and also the Turtle band, exchanged speeches, and agreed to the same number of hostages as the Senecas had done. They also promised to deliver up an Englishman charged with murdering and scalping a white man.

The formalities and business proceeded slowly, as usual, with Indians, who are loquacious in council in proportion as they are moody in conversation.

On the 11th, King Beaver presented six hostages and five deputies for the approval of the British commander. These he accepted, but told them that a chief of theirs, by the name of *Nettowhatways*, having refused to appear at the council, must be deposed, and another elected in his place. So thoroughly were they subdued that they obeyed this humiliating demand and chose another chief.

The Shawanese were dilatory, haughty, and sullen. But they appeared on the 12th with two chiefs, five head men, and forty warriors; and before all the tribes, *Red Hawke*, one of their orators, made the following speech. He spoke with pride, mingled with an ill-concealed submission, and said, "Brother, you will listen to us, your younger brothers; and as we discover something in your eyes that looks dissatisfaction with us, we now wipe away everything bad between us that you may clearly see.

"You have heard many bad stories of us; we clear your ears that you may hear.

"Brother, we saw you coming this road; you advanced with a tomahawk in your hand; but we, your younger brothers, took it out of your hands, and threw it up to God to dispose of as he pleases, by which means we hope to see it no more."

Red Hawke then produced a treaty made with the Penns, in 1701, and several letters from the proprietors. He then desired a renewal of those ancient friendships, promising that the prisoners now absent on the hunting expedition should be brought to Fort Pitt in the ensuing spring.

Colonel Boquet said in return, "that the speech would have been very agreeable if their actions had corresponded with their words," and that they had not kept their agreement, made with him at the Tuscarawas village a few days before. He asks them whether they will "forthwith deliver up all their prisoners and the Frenchmen living with them, and all negroes, without exception or evasion," and whether they will "give six hostages for the performance of this agreement and as security against further hostility?"

Benevissico replied that they agreed to all except as to the Frenchmen, whom, he remarked, were English subjects, over whom the Shawanese had no power.

During the winter, their hostages escaped from Fort Pitt; but when the tribes passed that way in the spring to meet Sir William Johnson, the prisoners they had promised were brought in.

On the 18th of November the army broke up its cantonment at the Whitewoman, and returned to Fort Pitt, which they reached on the 28th of the same month.

This expedition was conducted with so much skill and prudence that none of those frightful disasters that often result from Indian wars occurred.

The savages, although in great strength, found no opportunity to make an attack. No prisoners were taken, none died of sickness, and every man of the party returned except one, who was killed and scalped by an Indian when separated from camp. The Pennsylvania troops were under Lieutenant-Colonel Francis and Lieutenant-Colonel Clayton. Colonel Reid was next in command to Colonel Boquet.

The Provincial troops were discharged, and the regulars sent to garrison Fort Loudon, Fort Bedford, and Carlisle.

Colonel Boquet arrived at Philadelphia in January, and received a complimentary address from the legislature, and also from the House of Burgesses of Virginia. Before these resolutions reached England, the king promoted him to be a Brigadier-General. He was ordered to the command of the post of Mobile, and the next season died there.

Although I have extended this address to a great

length, there are a few incidents connected with the expedition that deserve to be heard.

A prisoner by the name of Small was one of a considerable party that belonged to a distant tribe. When they first heard of the invasion, the tribe determined to kill all the captives, and then collect their warriors and fall upon the troops. A Frenchman furnished them with some barrels of powder and ball, and they were preparing to execute their designs upon the prisoners, when a message arrived with the news that a peace was about to take place.

The tribe now proceeded towards Wakatomaka. When they arrived and there heard of the murder and scalping of the soldier, and that they were suspected of having committed the deed, the prisoners were again drawn out to be shot, and as the wretched creatures were standing in a field, preparing to die heroically, a second messenger arrived, stating that all suspicions of that tribe were removed, and that they would be received in council. This caused them to spare the prisoners, and to deliver them at camp alive.

The scene which took place when the captives arrived among their friends, I give in the words of Mr. Hutchins:

"Language indeed can but weakly describe the scene, one to which the poet or painter might have repaired to enrich the highest colorings of the variety of the human passions, the philosopher to find ample subject for the most serious reflection, and the man to exercise all the tender and sympathetic feelings of the soul."

There were to be seen fathers and mothers recognizing and clasping their once lost babes, husbands hanging round the necks of their newly-recovered wives, sisters and brothers unexpectedly meeting together after a long separation, scarcely able to speak the same language, or for some time to be sure that they were the children of the same parents. In all these interviews joy and rapture inexpressible were seen, while feelings of a very different nature were painted in the looks of others, flying from place to place in eager inquiries after relatives not found, trembling to receive an answer to questions; distracted with doubts, hopes, and fears on obtaining no account of those they sought for; or stiffened into living monuments of horror and woe on learning their unhappy fate.

The Indians, too, as if wholly forgetting their usual savageness, bore a capital part in heightening this most affecting scene. They delivered up their beloved captives with the utmost reluctance — shed torrents of tears over them — recommending them to the care and protection of the commanding officer.

Their regard to them continued all the while they remained in camp. They visited them from day to day, brought them what corn, skins, horses, and other matters had been bestowed upon them while in their families, accompanied with other presents, and all the marks of the most sincere and tender affection. Nay, they did not stop here; but when the army marched, some of the Indians solicited and obtained permission to accompany their former captives to Fort Pitt, and employed themselves in hunting and bringing provisions for them on the way. A young Mingo carried this still farther, and gave an instance of love which would make a figure even in romance. A young woman of Virginia was among the captives, to whom he had formed so strong an attachment as to call her his wife. Against all the remonstrances of the imminent danger to which he exposed himself by approaching the frontier, he persisted in following her, at the risk of being killed by the surviving relatives of many unfortunate persons who had been taken captives or scalped by those of his nation.

Among the captives a woman was brought into camp at Muskingum, with a babe about three months old at the breast. One of the Virginia volunteers soon knew her to be his wife! She had been taken by the Indians about six months before. He flew with her to his tent, and clothed her and his child with proper apparel. But their joy, after the first transports, was soon dampened by the reflection that another dear child about two years old, taken with the mother, had been separated from her, and was still missing, although many children had been brought in.

A few days afterwards, a number of other persons were brought in, among them was several children. The woman was sent for, and one supposed to be hers was produced to her. At first sight she was not certain; but viewing the child with great earnestness she soon recollected its features, and was so overcome with joy, that, forgetting her sucking child, she dropt it from her arms, and catching up the new found child, in ecstasy, pressed it to her breast, and bursting into tears carried it off, unable to speak for joy. The father, rising up with the babe she had let fall, followed her in no less transport and affection.

But it must not be deemed that there were not some even grown persons who showed an unwillingness to return. The Shawanese were obliged to bind some of their prisoners, and force them along to the camp; and some women, who had been delivered up, afterwards found means to escape, and went back to the Indian tribes. Some, who could not make their escape, clung to their savage acquaintances at parting, and continued many days in bitter lamentations, even refusing sustenance.

TWO MONTHS IN THE COPPER REGIONS.

[National Magazine, February, 1846.]

It was on the 14th day of August, 1845, that our party went on board a light and well built yawl, of about four tons, moored in the still water above the rapids of the St. Mary's River. We were venturing upon an experiment. We could not learn that such a craft had ever put forth alone upon the waters of Lake Superior, and our intention was to follow the south coast as far as the season would permit. For hundreds of years this lake had been navigated by the bark canoe, and parties were setting off every day for Copper Harbor, La Pointe, and other remote points, in these apparently frail vessels, but which the experience of centuries had demonstrated to be the safest conveyance known. The Mackinaw boats had long traversed these shores, transporting goods to the Fur Company's posts, and returning with furs.

These long, narrow, flat-bottom boats, carry a heavy burden, go well before the wind, and are easily drawn ashore. The bark canoe, like the Mackinaw boat, has no keel, and the safety of both consists in being able to make a harbor of every sand beach, in case of a storm. The expert voyageur has a kind of second sight in regard to weather, smelling a storm while it is yet a great way off. It is only when a great saving may be made, and the weather is perfectly fair, that he ventures to leave the vicinity of the shore, and cross from point to point in the open sea. These passages are called "traverses;" and such is the suddenness with which storms arise, that a traverse of ten or fifteen miles, even in fair weather, and while every indication is favorable, is regarded as a hazardous operation. Some daring boatmen make them of thirty miles.

Of course, the birch canoe and the Mackinaw boat, being without keels, cannot sail upon the wind. Our yawl, with a keel of four inches, having nine men and a ton of provisions aboard, sank about sixteen inches in the water. She was provided with a cotton square sail, containing forty square yards, and had row locks for six oars. How she would row, how she would sail, and how she would brave the storm, we could only surmise, and the surmises were rather against the little vessel.

The portage over which goods now pass, from the level of Lake Huron to that of Lake Superior, is a flat, wet, marshy piece of land, about three-fourths of a mile across. To the westward, the country appears to be low and swampy as far as the view extends, which however is limited by the thick timber, principally spruce, pine, white cedar, birch, and hemlock. But a walk of one mile in that direction brought me to a low eminence rising out of a cedar swamp, composed of masses of rolled granite and other primitive rocks, in size from a small pebble to a diameter of ten feet. The timber among them had been lately blackened by a raging fire. The trunks of these charred trees, some standing erect, some leaning against others, and many prostrate on the rocks, contrasted hideously with the white and nakedness of those immense granite boulders which covered the surface.

On the north and east, in the province of Canada, a high range of mountains extends in each direction out of sight. They were first visible at the head of

St. Joseph's Island, having the jagged outline of trap-rocks. The view from the low ground on the American side, towards the high land across the river, is extensive and gratifying. In front is the river, a mile broad, and the rapids. At the opposite shore, the establishment of the Hudson's Bay Company, half commercial, half military, with a stockade and white houses. For several miles down the river there are houses on the bank, and farms extending back, at irregular distances up the mountains. Here the traders, voyagers, missionaries, factors, Indian agents, and Indians, reside promiscuously—such is the foreground of the view. Behind and beyond rise the mountain ranges, in that pure atmosphere perfectly distinct at the distance of twenty miles.

Our tents were struck at 7 o'clock, A. M., and the journey began. There were other parties scattered about the open space at the warehouse; some had regular tents, some sheltered themselves under a broad piece of India rubber cloth, stretched over a pole like the roof a house. One party had a conical tent, with an upright pole in the centre, the canvass spread out around the foot; and another, in default of other covering, lay snoring under a cotton bed-tick, stretched across the bushes. A party of surveyors were encamped near the landing, from a cruise of three months in the interior. This party had run a tier of townships, from Mackinaw northward, into sections of one mile square. These men encamped a few days at this place to recruit their tattered garments, of which only the shreds and fragments remained. In enterprises of this sort, it is only by physical energy, and great powers of endurance, that the contractor can realise anything from the prices .allowed by Government for its original surveys. They provision themselves by carrying all on their backs, from the depots on the shore. The thickets through which they pursue their work, week after week, and month after month, would be declared absolutely impracticable to a person not trained in that school. No beast of burden could pass without bridges, even in case a pathway should be cut through the matted evergreens that cover the ground. To make a path for a horse or mule would consume more time and labor per mile than the survey itself. There is a hardy class of Frenchmen and half-breeds, cousin-german to the Canadian voyageur, called "packers;" they were bred in the service of the Fur Companies, to carry goods from the nearest landing to the trading post, and return with a pack of furs. The surveyors found these packers indispensable to their operations. They will carry from fifty to seventy pounds, and can travel alone in the recesses of the forest without fear of losing their way.

They are patient, cheerful, and obedient; in fact, they are on land what the voyageur is upon the water. His capacity for food corresponds with his ability to endure fatigue, and his great care is to secure it in sufficient quantity. He makes, with a little instruction, an excellent axe-man and chain-man. If circumstances prevent a return to the camp, or the rendezvous, he can lie down at the foot of a tree, sleep till daybreak, and resume his tramp without complaint.

The party which joined our encampment here was a subject for Catlin, the Indian sketcher. More hale, hearty, and jovial fellows, never broke into the limits of civilization. The northern atmosphere had tinged their cheeks with red, they were all young and active men, glowing with that high animal life, that extreme buoyancy of spirits, which is a stranger to the inhabitants of cities, to those who toss upon feather beds, and live upon soups and comfits.

This rugged company, full of fun and frolic, with beards of three months' cultivation, red flannel shirts, and fustian trowsers in shreds, white beaver hats, less the border, some in shoes, some in moccasins, and some in boots, from all of which various toes were looking out, surprised even the worthy burgher of the Sault. The Sault St. Marie has been a trad ing post more than two hundred years. The goo Catholics, *Ramboult* and *Jouges*, preached repentanc to the *Nodowessies*, or Sioux, on this spot, in 1641 whom the French traders immediately followed. Her it may be said the borders of civilization have bee fixed for two centuries. In consequence, a mixe race has arisen, neither the representatives of refine ment nor of barbarism, but of a medium state. It ma well be supposed, that a band of jolly fellows, habite as we have described these hardy surveyors, axe-men chain-men, and packers, would not attract here tha attention which they would in New York, or in Lon don. But they appeared to be objects of no litle in terest and curiosity to the worthy inhabitants of th Sault, especially as some of them were so disfigure that their old friends did not recognize them.

Looking back from the water, upon the collectio of tents, and lodges, we had a view of the group a one glance, and the scene from the new point of ob servation suggested ideas that had not presente themselves while we formed a part of it. Aroun some of the camp-fires were gentlemen from the At lantic shores, with genteel caps and surtouts, shiver ing in the raw wind of the morning. Poor fellows impelled by the hope of wealth to be found in th copper region, they had rushed at steamboat rates t the extremity of navigation, of taverns, and perma nent habitations.

The reality of copper exploration had now com

aenced. A night of drizzling rain and fog had been ›assed, in a cold tent, on wet ground. Among them vere seated voyageurs and half-breeds, as happy as a ›lenty of grub could make them. The raw wind was ıo annoyance to them, so long as there was a flint ›nd steel to start a fire, and a plentiful stock of pro-isions. Between the cap and surtout, and the flan-ıel shirt and canvass trousers, was every grade of men epresented by a grade of habiliments.

In front of this motley collection of persons and hings lay the frame of a large schooner, on which ifty workmen were laying the plank — all its timbers ›nd lumber brought from the lower lakes; and in the ›pen level space beyond, along a track cleared through he swamp, stood the spars of a vessel, advancing on olid land towards the basin above the falls. The la-›or and expense of bringing vessels overland, or the imber to construct them with, is unavoidable. As 'ar as known, there is not ship timber enough on Lake Superior to build a schooner.

The rock which causes the rapids is a close, fine-grained red sandstone, in thin layers, pitching to the ıorthward. There has been much diversity of opin-on among geologists, about the geological position of his rock. As I proceed, I shall again notice this 'ock, and its analogy, which occupies almost the en-ire southern coast of the lake.

The first *principal meridian* of the United States surveys comes out on the waters of St Marie's, at the ship yard, just above the rapids. This is a *true meridian*, run with great care from the *base line*, which is about twelve miles north of Detroit. The first meridian is about thirty miles west of Detroit, and, passing up through the peninsula of Michigan, crosses the straits near Mackinaw. By the Govern-ment. system of rectangular co-ordinates, referred always to a given base and meridian, an observer knows his exact position, wherever he may be, in the surveyed portions of the United States. Every township is six miles square, every section one mile square, every quarter section half a mile square. Every section *corner* has permanent marks on some adjacent tree which gives the situation of that corner from its proper base and meridian. I make this explanation to give light upon terms that I shall use hereafter. In traversing the American shore of Lake Superior, we found as far as the Porcupine Mountains, west of the Ontonagon, that the surveyors had preceded us. During the present and the past year they had extended the township lines to this distance along the coast, and for a part of this distance had sub-divided the townships into sections. The surveys had been carried to different distances, interior. From the base near Detroit, numbering northward, St. Marie's is in township No. 47 north, range No. 1 east. But our point of embarkation was on the west side of the meridian, in town 47 north, range 1 west, or 282 miles north of the base line.

We are now fairly under way, and shall be able to keep our reckoning. The river expands, as we ascend against a very gentle current, the shores are low and swampy, or sandy, and covered with stinted pines. In an hour and a half, so easily did our boat row, we were at "Point Aux Pins," on the British side. At ten o'clock, we were on shore at "Gros Cap," looking up a spar, and clambering the red granite ridge, which here projects towards the American shore —the extremity of that range of mountains in view from the rapids to the eastward. From the height of 500 feet, we could see the continuation of this range westward into Michigan, until its summits were lost in the mist. The eastern extremity of the American range is "Point Iroquois," nearly opposite "Gros Cap," where the Chippeways, by their ancestors, fought a great battle with the Iroquois, long before the French came into these waters. The range is called the "Tequamenon Mountains," overlooking for some twenty miles a deep bay, known as the Tequamenon Bay. The waters about "Gros Cap" are so clear that the bottom is seen from fifty to sixty feet below the surface.

Before leaving this inhospitable crag, we set fire to a windfall of about two years of age, and consequently in a fine state for a conflagration. This was not done through any republican contempt of the British Queen, or her territory, but from pure benevolence towards subsequent travellers exploring "Gros Cap." It lay between the ridge and the bay, in a swamp so thickly covered with prostrate trees that one might go a quarter of a mile on them without touching the ground, unless an unlucky misstep should precipitate him into the mud beneath. At 1 o'clock, we were at "Isle Parisien," a low island, five miles long, cooking a dinner and procuring a better spar.

We succeeded here so well in fitting our sail, that the traverse of fifteen miles to "White-fish Point," ordinarily a hazardous voyage, was safely and pleasantly made, a little after dark, and the wind, though light, being still fair, we ran into the lake without landing, and made along the shore. We were now upon the largest body of fresh water on the globe, called by the Indians *Kitche-gomig;* by the French, *Superieur*, or Upper, and corrupted by the English into "Superior."

The moon shone dimly through a heavy sky, the water was merely ruffled by a warm southern breeze, and in the distance the flame of the burning windfall shone conspicuously above the mountains. On the Michigan side several large tracts of burning timber were seei on the hills at the head of Tequamenon Bay. It wa determined to proceed as long as the wind continue favorable, but in a short time it failed altogether, an we went ashore at half-past eleven and encamped The ground here lay in a series of low sand ridges with scattered pines. Distance from the Sault, forty five miles.

At sunrise everything was on board, and the sai spread before a fair wind. Along the beach the sur has piled a ridge of water-washed granitic gravel, fiv to six feet high, the deep water holding out quite t the shore. In coasting in an open boat, the travelle must resign all hope of regularity of hours, of meals and of sleep. His sovereign is the weather: whe that is calm, he may proceed with the "white-as breeze," as the sailors say: when the wind is ahead he can take his ease—provided he is safe on shore but when it is fair, he must always be before it. Th prevailing winds along this shore are from the wes at this season; and consequently they are ahead a you go up the lake.

Breakfast on board, upon cold beans, cold pork and hard bread.

Towards evening the wind came so strong ahea as to oblige us to put into the mouth of "Two-Heart" River, a stream sufficiently deep to float a large ves sel inside the bar, but not deep enough to carry th yawl with her load. Of the streams discharging int the lake from the south, only two or three are know with open mouths. At most of them it was necessar to lighten the boat and haul her over, with about th same labor and discomfort as though there was n channel; but, once inside, a quiet harbor was alway found. These mouths are so completely sealed up and concealed by sand ridges, that persons may pas hem within ten rods of shore, and not discover that a reek is there.

The shore is composed of low, monotonous sand idges, with stinted pines. The bluff is from fifty to ighty feet high, presenting a stratified edge of sand, iclined gently to the east, not exceeding ten feet in mile. The ridges run from the interior nearly per-endicular to the direction of the shore.

We passed several fishing huts, now deserted, with lenty of empty salt barrels and fish-scales scattered ound.

A little east of the mouth of the creek we observed, sailing up, several picketed enclosures among the nes on a beautiful ridge. They were Indian graves, ius strongly guarded to keep out the beasts of prey. here are those who doubt whether the Indian is sus-ptible to the delights of taste—whether he enjoys a ight morning, a clear and moonlit night, a moun-in, a vale, or a beautiful river. Was it mere acci-nt that placed this burying-ground upon so enchant-g a spot? The lake is about forty rods distant in ont, and about as many feet below the site of the aves. Through the open trees you see its waters, plainly as if there was no intervening timber, while e shade of their branches is perpetual upon the spot. ven the lowest ripple on the beach reaches the ear distinctly as the angriest roar of the waves. Eve- breath of air that moves to and from the lake, the ening and the morning breeze as well as the north-n tempest, plays audibly upon the long and ever-een leaves of these ancient pines. At the head of ch grave is a flat shingle or board, with emblems inted in red, or rudely carved with a knife. On e there are three red crosses, and two human fig-es, representing a man and a woman (doubtless a sband and wife), with clasped hands. On the reverse, a bear, probably the sign or totem of the deceased. On the top, three eagle quills. Some have crosses, indicating that a good Catholic sleeps below.

At an early hour on the morning of the 16th we got out of "Two-Heart" River through a light sea, determined to try the "ash breeze" against the west wind; but after a couple of miles hard rowing, the regular breeze prevailed: we could no longer make headway, and put about.

Notwithstanding the sand-flies and mosquitoes, i was comfortable to lie down once more upon the gree grass and fragrant wintergreens of that shore. Th weather was warm and hazy. Some wandered throug the sand-hills and swamps; some, wrapped in blankets as a defence against the flies, sought in vain fo sleep; others, with the fish-hook and artificial fly rowed up the creek in pursuit of speckled trout. A good dinner and supper of these fish was the result o the expedition.

At 8 P. M., the wind became more favorable, an the boat was headed up the coast. At 10 the weather became thick, and running ashore at random, w had the first trial at hauling our craft out of the wate by main force. She proved to be as easily handled o land as a Mackinaw of the same capacity, only requirin more care. For a camp we turned her over, one gunwale resting on the sand, about thirty feet from th surf, the other set upon sticks after the fashion of trap. Under this we all crawled, spread our blanket and *some* of the party went to sleep.

Mr. J. R. Dorr, of Detroit, the principal of th expedition, had seen something of this kind of lif Mr. D. P. Bushnell, of the same place, had long bee Indian agent at La Pointe, and was, of course, famiiar with the country and this mode of travelling. An other gentleman, well known on the lakes for his w and vivacity, qualities that generally attend an excitable temperament, not being accustomed to tents, boats and camps, found it rather uncomfortable. The sand, so soft and yielding to the foot, was as hard as a rock to the bones. The grinding of the gravel, thrown incessantly about by the waves, gave out a grating sound that had no tendency to soothe a man to rest, especially one who had been accustomed to the quiet of the third story of a boarding-house. Besides, there was some chance of the props giving out, and the trap being sprung upon the legs, arms, and bodies projecting from beneath. *Mike*, an old soldier who officiated as cook; *Martin*, a sailor just from the whaling grounds of the north-west coast; Charley, a giant from the low countries, and Patrick, the other hand, seemed to pay no attention to the hard bed, the cold wind, the noisy waves or to the doubtful props. A sprightly young clerk of the company, fresh from the counter, though swollen and tormented by the poison of the sand-flies, took the matter like a veteran, and slept like an opium-eater.

About noon the next day we passed the "Grand Marais," a bay forty miles from White-Fish Point, with six feet water on the bar, and a fine harbor.

Two men had left St. Marie's the day before we did, in a small but neat and clinker-built boat, with two masts and a wide keel. They were wholly unacquainted with the difficulties that lay before them; yet one of them, by the name of *Axtel*, had been exposed in the same boat forty-eight hours to the fury of a Lake Michigan storm, and therefore felt a confidence in fate. Neither of them had been on Lake Superior, and therefore knew little of its harbors, rocks and storms. Their supplies were salt pork and bread, their furniture a camp-kettle.

Passing Grand Marais, before a smart breeze, we

saw their fire in the harbor, and shortly their sail coming up astern. Here the low, regular, dreary shore of sand suddenly changed to a lofty wall of the same material, rising from the water's edge as steep as it will lie, to the height of 400 feet. For twenty miles back, there had been seen near the water's edge a stratum of *pebbles*, inclined with the sandy stratum above it, to the eastward. Now the strata of sand rests on a bed of *clay*, with the same inclination, but only a few feet in the mile. The *Grand Sable* struck us with more force, because of the sudden transition from a low uninteresting shore, to a bold, lofty, regular scarp, four times the height of the tallest trees. But there were upon this Sable no trees or other vegetation, either on the face towards the lake, which was nearly perpendicular, or upon the summit; all was one bleak pile of sand, yet so clear, so regularly stratified and so beautifully variegated by colors, white and red, that the prospect was not dreary, but rather sublime. Imagine a straight wall of pure sand, four miles long, and four hundred feet high; the base lashed by a rough sea, its top enveloped in a heavy mist, through which rounded hillocks of white wind-blown drift occasionally rise, as the eye reaches, mile after mile, over the country behind. To me this sight was more grand and curious than the Pictured Rocks. Whence came this mass of sand? Its upper portion has apparently been moved about by winds, its lower portions appear to be too solid to be thus moved. Was it not in remote ages like the low sands we have passed, but extending much farther into the lake? A prevailing north wind, with sufficient force to move the sand at the surface, would overcome vegetation, and, like the current of a river, transport the particles incessantly in one direction. By this means the sand would pile higher and higher, and the lake, alvays encroaching at the foot, would increase the ieight of the bluff shore.

The "Sable" overlies on the west, a variegated sand rock, coarse-grained, and easily broken, pitching slightly to the eastward. This is the first rock west of White-fish Point. The stratification is imperfect, the color an irregular mixture of gray and red.

Turning one of the rocky points west of the "Grand Sable," a stiff gale from the west put an end to further progress, and gave warning of a storm. The only expedient in such an emergency is to beach the boat, and draw her out of the reach of the waves. It is an operation not always agreeable, because while loaded she cannot be run upon dry ground, and to be unloaded the goods must be taken through the water to the shore. On this occasion the wetting process had been gone through with two hours before, duriug a heavy fall of rain.

Our baggage was scarcely safe on land when the wind blew furiously, and our two friends in the sail-boat appeared, endeavoring to make the shore, as the sea had risen so much that a landing was at this moment not only uncomfortable, but a little hazardous.

As the storm increased, our fires began to burn brightly. Near the boats was a little dell sheltered by a low ridge of sand, where our tents were pitched, and all made dry and comfortable, while the gale heightened into a tempest.

On the next day progress was impracticable, and being well provided, we determined to give an entertainment. Our friends were invited at 1, P. M. We had bean soup, boiled ham, tea and coffee, bread, and *pickles*. The quantity consumed probably exceeded that of ordinary dinners, as much as it does at the annual meals of the Aldermen of New York and London. As to style, there were tin cups and pewter platters, knives and spoons. For tables, there were the knees of the guests and a spare box; for seats, camp-stools and bundles. The entertainment continued with great glee about two hours, and passed off with as much sociability and mirth as though it had been given at the Astor.

After the first hour had been spent in the enjoyment of this cheer, our guests began to refuse dishes, by way of politeness; but the ex-Indian agent put all such hesitation aside, by relating what he had done and seen in the Indian country. There was one example of an Indian eating half a bushel of wild rice at a meal. Another, of a half-breed, who was sent out to bring in a deer that had been killed some miles from the post. The half-breed lost his way, and slept in the woods one night. The next day in the afternoon he came in without the deer. He was asked where he had left it. "Ugh! eat him—do you s'pose a man is to starve?"

One thing is certain—in this high latitude, with its pure and healthy climate, where the enervating effects of heat upon digestion are unknown, men may eat with impunity what would be fatal to them at the south.

In commemoration of the feast, a little trout brook which empties there was named "Pickle Creek," and the names of the party neatly carved on a neighboring birch.

One of our guests is the son of a former sheriff in Canada, who made the journey from St. Marie's to Fort William by land, in the winter of 1816. The object of this trip, through a region so rough and forbidding in the severity of the cold season, was the execution of a warrant upon Lord Selkirk, then in possession of that post. Fort William is situated about the middle of the north shore, nearly opposite the east end of Isle Royal. The warrant was issued from the King's Bench, and had reference to some of those acts of violence that occurred between the "Hudson's Bay Company" and the North-west Company." The sheriff, whose name was *Smith*, at last reached the fort, with ten men. Selkirk professed to hold and to fight under the ancient chartered rights of his ancestors, and when Smith presented his authority for the arrest, Selkirk fell back on his charter. Smith offered the authority of the King's Bench; Selkirk claimed to be outside of all civil jurisdiction, and replied: "If you do not believe in my charter, here is my authority," pointing to about fifty men who were ready to do battle in such emergencies. He continued: "Instead of my being your prisoner, you are mine. I will treat you and your men well, yet you must take quarters in the block-house till I leave here."

Accordingly the sheriff was obliged to remain in custody about five months, until the opening of the season.

The timber about Pickle Creek is black and white birch, a few stinted white maples, white and yellow pine, mountain ash, spruce, balsam of fir, balsam of spruce, white cedar, and hemlock, none of it large enough to be valuable.

The next morning at 4, with a fair wind, we were on the water, having Grand Island in sight at daybreak. This island is high and bold, like the Pictured Rocks, which lie on the mainland opposite. It bears sugar-maple in profusion, and has one family (that of Mr. Williams) residing upon it; he is a thrifty farmer and trader. The variegated sandstone, as well as I could determine, here plunges to the west, and passes under the strata which compose the Pictured Rocks. The lamented Dr. Houghton regarded the red or variegated sandstone of Lake Superior as older than the "old red sandstone." The Pictured Rock stratum he considered the equivalent of the "Potsdam sandstone" of the New York Reports. This rock comes to the shore, about twenty miles in length, and has a thickness of at least five hundred feet. Grand Island is an outlier on the north.

The following is a section from the water's edge upward, taken by the eye, at the highest point, which, according to Captain Bayfield, is 300 feet.

Geological Section of the Pictured Rocks.

	THICKNESS.
1st—Soft conglomerate, ...	10 to 20 feet.
2d—Compact sandstone, ...	60 to 80 "
3d—Gray sand rock (soft and wearing), .	30 to 50 "
4th—Yellowish green sand rock, containing sulphur, sand easily dissolved, . . .	80 to 100 "
5th—Soft and imperfectly stratified sand rock, with septaria of sandstone, and ironstone, ...	20 to 50 "
6th—Drift and sand to surface, . . .	10 to 50 "

It will now be readily seen how the perpendicular faces of rock are caused, which have given this passage such a frightful aspect. Vertical walls of smooth gray rock, 200 to 300 feet high, passing to unknown depths beneath the surface; in places worn into large caverns, in others, colored in fantastic yet grim figures, half real and half imaginary, yellow, green, and black; shapes neither animal, nor in the likeness of anything else that is natural, but so near the natural as to give rise to the idea of monsters, griffins, and genii. Such are the Pictured Rocks, before which the Indian thinks of his Manitou, and the Frenchman crosses himself with profound reverence.

The soft conglomerate (No. 1) yields to the incessant wear of the wave, which, rolling in from deep water, strikes with great power. When the undermining process has extended a few yards, the hard stratum next above falls, and with it the superincumbent mass. Much of this dissolves away in time, leaving the fragments of No. 2 visible in the clear water, in great blocks, at various depths beneath the surface. The colors are furnished by the dripping solutions of iron, in the state of oxyde, carbonate, and sulphate; by moss growing upon the face of the rocks, and probably by the green carbonate of copper. The niches, caves, and angles, follow naturally from a rock of different degrees of hardness, acted upon by the same disintegrating force. At the mouth of a creek, where the trail from "Bay De Noquet" (called *Badenock*), on Lake Michigan, strikes this lake, there is a hard silicious slate, approaching to flint, dark in color, and imperfectly stratified. This bed, which appears to be limited, lies low, near the water.

Passing these dreaded rocks, the principal harbor of Grand Island and the farm of Mr. Williams came in view. For refuge in bad weather, this island must in future time be of great advantage to vessels. It has several large and deep harbors, and of itself forms a good lee, in almost all weather. On the mainland, opposite Mr. Williams', is a solitary cabin, the agency of the American Fur Company.

Between Grand Island on the west, and the shore at Train River Point, there are two low islands that appear to be formed of the red sandstone. At the point, this rock forms the shore, and has a rapid dip to the eastward, say 150 feet in the mile; evidently running under the Pictured Rocks, and therefore an older formation. Here it enclosed occasional pebbles of quartz, agates, and fine-grained sandstone.

The wind which had been fair all day, on turning the point came strong ahead, against which we had hard pulling about five miles, to the mouth of Train

River. Our craft proved to be a fast sailer, easily beating the little clinker of our friends before the wind; but those dauntless fellows did not rest until, at the end of the day, they drew into the same harbor with us. Train River, like many others, has deep water inside, but only a few inches at the entrance. Wherever we set foot on shore, the remains of previous travellers were seen. Here the poles of many Indian lodges were standing, and the bones of a bear lay around, indicative of a feast. There were also dwarf cherries and whortleberries.

Passing out of the bay in the morning, a range of mountains was visible, the ends presenting themselves near at hand, and the principal range extending westward, towards Chocolate River. From the outline I conjecture that they are composed of primitive rocks. At the shore the strata are still the variegated sandstone, very much tilted, with thin beds of shale interstratified; apparent dip, to the northward. Making a long traverse from Train River Bay, at 5 P. M., we entered a magnificent harbor between projecting points of granite rocks, and coasting along inside some islands soon saw that there was a very safe and spacious shelter for shipping still further inland, accessible in any wind, with deep and quiet water inside. This bay is sometimes called Presque Isle. It commences about two miles north of the mouth of "Riviere des Morts," six or seven miles north-west of Chocolate River, and extends to Granite Point.

Mr. Dorr being quite ill, our party remained a day. The boat anchored in a quiet nook of the harbor. Granite rocks were projecting on all sides, through the red sandstone, scorched and whitened at the points of contact. In the rear were seen rugged mountains, covered with evergreens. This was regarded as the commencement of the copper region.

Accordingly, myself and Martin sallied forth in the morning to spy out the mineral wealth of the spot. On the south point of the bay, to our great satisfaction, we discovered a piece of green carbonate, about the size of a pea, in a hard green stone-trap; but a little further on found, also, evidences of a prior occupation, in a log cabin covered with birch bark, a small patch of chopped land, and a pen made of poles, which enclosed two or three hills of potatoes, and some stalks of green peas. Pursuing our way along the shore to Dead Men's River, we found a permanent fishing establishment and two comfortable houses, now deserted and locked up.

The country adjacent for two or three miles is low and swampy, with sand ridges between the swales, and at the mouth of the river heaps of granite rocks. It was soon evident that the surveyors had been this way, and that very recently. At the south point of the bay was a stake, on the dividing line between sections Nos. 1 and 2, town 48 north, range 25 west; showing that we were one town, or six miles, north of St. Mary's, and twenty-five towns, or 150 miles west.

In making the traverse from Train River Bay to Presque Isle Bay, a singular object was visible to the sight, long before the shore opposite to it came in sight. Under the effect of refraction, it rose and fell, dilated and contracted, changing continually from a tall spire to a flat belt of land. By the glass, it was seen to be almost destitute of trees, and Mr. Bushnell began to regard it as one of the peaks of Point Kewena. There is no map of this lake upon which a navigator can rely, except a British one from the survey of Captain Bayfield (Royal Navy), made about twenty years ago. We had what purported to be a copy, but soon found that it was not a true one. We could neither recognize from it the harbor, the points, nor the rivers where we were. At Chocolate River the coast, from a westerly course, makes almost a right angle to the northward; but at that time whether we were at Chocolate or Dead Men's River we could not tell.

The isolated object seen in the north proved to be the "Granite Rock," situated about ten miles from shore, fifty to eighty feet in height, and a few acres in extent. Along this shore, huge masses of this *recen* granite rise through the water, and may be seen in it clear depths. From the section stake just mentione the Granite Rock bears north ten or twelve miles distant. It must not be confounded with "Stanard's Rock," which is in the track of vessels from Poin Kewena to St. Mary's, thirty miles from land. Tha these granite rocks are more recent than the sedimentary sandstone which rests upon them is eviden from observation. The igneous rocks have *protrude through* the sandstone, distorting and breaking up th strata. If the red or variegated sandstone had bee deposited *after* the upheaval, this disturbance woul not have been visible, nor would there have been see the discoloration and semi-vitrification at the junctio or contact of the two formations.

The mountainous country which here comes quit to the lake extends, in east and west ranges, beyon the sources of the Huron River and Kewena Bay, an appears to have been formed by the same volcanic ef fort. The spacious and beautiful harbor where we la is formed by four granitic islands, three of them no connected with the shore by sand bars, forming as many "Presque Isles."

Our next day's sail ended at a small creek, by th voyageurs called *Cypress* River, from the adjacer forest of cypress timber, as it is called. This tree an evergreen, with rough bark resembling a tamaracl but the leaves are more like the hemlock. At fifteen miles from Presque Isle Harbor the shore made again to the westward, the sandstone bluff being more elevated and perpendicular; its strata somewhat rolling, but the general dip appeared to be westward. The face of the rock was from fifteen to eighty feet above the water. The knobs of Point Kewena were now distinctly in sight, from forty to fifty miles distant in the north. Mr. Dorr being quite sick with a bilious fever, we determined to make a long traverse on the next day, across the bay to Traverse Island, and thence, with all dispatch, to Copper Harbor. But after putting out in the face of a stiff breeze early in the day, we found it impossible to weather the next point, and returned to camp.

The river, called the St. John's by us, is known to the French as the "Chien-Jaun," or Yellow Dog River, corrupted in the first instance to "Shannejone," and thence to St. John. It is on the map laid down as about thirty miles long. In this country the word creek is never used, but the French term "*rivieere*" is applied to all its streams, which is Anglicised *river*. Being now wind-bound for the day, I took our trusty and intelligent whaler, *Martin*, who had already shown himself a good woodsman, as well as a first-rate sailor, and followed the creek into the interior. At the end of two miles of still, deep water, our canoe was obstructed by flood-wood, and at three-fourths of a mile further, by a fall of eight or ten feet over sand rock.

Above the fall was a beautiful lake, overlooked by granitic mountains on the west, with an opening at the south. This led us to a second lake, and this to a third—strictly speaking only branches of the same water—in all about four miles long. On the east and south were gentle ridges, sustaining the first valuable pines we had seen; on the west, lofty hills. In the low grounds, at the water level, were thousands of large white cedars, forming a perfect *abattis*, or barrier, against our progress. There were pheasants and ducks in abundance, red squirrels, and whortleberries. On the whole, there was present so much of the New England scenery and productions that I have written on my sketch of these ponds the name of "New England Lakes."

On the succeeding day, the wind being still adverse to a direct passage to Copper Harbor, we thought only of proceeding along the coast to the westward, and reached the mouth of Huron River in a few hours. The health of our invalid having improved, we hauled the boat over the sand-bar at the mouth of this river, and finding deep and wide water, ascended about two miles and encamped. The reports of other exploring parties were highly favorable to the Huron region, as a mineral location, but after expending two days of rainy weather, in the mountains between the Little and the Big Huron, and finding the signs of valuable copper not promising, we set forward for the "Anse."

During our stay we had made an excursion by water into a bay about fifteen miles deep, called after the adjacent islands and river, Huron Bay. The shores are low, and the extremity or head swampy, and filled with a labyrinth of wet islands, covered with white cedar. On the south, the Huron range overlooks the bay, at a height of 500 to 600 feet. This inlet is in the form of a pocket gathered at the middle, and if necessary, though shallow, would accommodate a great number of vessels. When we were fairly at the bottom of the pocket, the wind came square in, and preventing our departure that night, we were under the necessity of encamping without blankets in a lodge lately occupied by the surveyors.

A lodge is a temporary habitation erected by those who have no tents, to be occupied for the night, or for some days if the weather is bad. It is made of evergreen boughs, pine, hemlock, or balsam, cut short. the frame-work consists of two crotches, and a pole between them. On the side towards the wind poles are laid like rafters, one end on the ground, the other on the cross-pole in the crotches. On these the small brush is laid like shingles, beginning at the ground, and each course overlapping the last. The ends are stopped in the same way, and the fire built in front. They serve to keep off the dew, snow, and wind, but are of little avail in heavy rains.

The promontory between Huron Bay and Kewena Bay is called "Point Obang," a corruption of "Point Abaye." It is a low, flat tract of land, which bears some sugar-maple, and has a good soil capable of cultivation. The range line, between ranges 29 and 30 west, comes to the lake a short distance west of the mouth of Huron River. The north-west corner of section 18, T. 52 N., R. 29 W., is about a mile from the shore, showing a progress to the westward of St. Marie's of twenty-nine towns, and to the northward five towns.

About six miles from the shore is a collection of granitic islands, called the Huron Islands, inhabited by rabbits in great numbers. Soon after casting loose from the islands, our fitful breeze again settled into the west, compelling us to land and anchor the boat in an open roadstead, where she tumbled and pitched all night and all the next day, our faithful whaler sleeping on board. In the evening a calm enabled us to work with oars, and to reach the mission at the "Anse" about daybreak.

This term is the French for a small bay, and is used to designate the place as well as the head or extremity of Kewena Bay. Here the *Abbe Mesnard* preached to the Sioux, in 1660, and impelled by the missionary spirit proceeded towards "Chegoimegon," the modern La Pointe. He is said to have perished in the wilds of Portage Lake, for he was seen no more.

There is yet a Catholic mission on the north side of the bay, which, with its collection of log cabins, and chapel, presents at a distance a very pretty view. On the south side is the Fur Company's agency, now comparatively desolate, and the Methodist mission for the Chippeways. Dr. Johnson, the carpenter, and Mr. Brockway, the blacksmith and farmer of this mission, showed our party great kindness, which is more to be considered, when it is known that the spirit of copper speculation had attracted many people to the country, all of whom received the good offices of the etsablishment.

The mission farm produces good grass, very heavy crops of potatoes and turnips, good oats, barley, and rye. They are now trying the wheat crop, with little doubt of success.

Those who have spent the winter here do not complain of its severity, although snow lies from one to four feet deep, from December till May. The bay furnishes inexhaustible supplies of white fish that are taken almost the entire year. Every night, except Sunday, the water is dotted with the canoes of the squaws and Indians, planting their gill nets; and again at daylight in the morning these female fishermen are seen overhauling the net for their morning meal. The two missions appear to divide the band about equally. At this moment the principal portion of both flocks are absent at La Pointe, receiving their annuities, each under the watchful care of their respective pastors.

From the Anse to the mouth of the Ontonagon, direct by land, is a very practicable route for a road, the distance about forty-five miles. It is from this place, also, that the winter trail to Greeen Bay leads off to the southward, and which must always be the approach from the States by land. To reach the Ontonagon by water, the distance is about 160 miles, following the shore around point Kewena. But about twelve miles from the Catholic mission there is a river, called the Portage River, that communicates with the Portage Lakes, which extend across the base of Point Kewena, to within one mile and a half of the northern shore. For bark canoes and light craft this portage is practicable, and is usually made. About sixty miles of navigation is thus avoided.

Having feasted a couple of days upon the good things of the Anse,—to wit: potatoes, turnips, sweet milk, and fresh bread,—we departed for Copper Harbor, and arrived there in two days. The sand rock of the south shore of Kewena Bay continued around on the northern side to "Bay de Gris." A little beyond this, a different rock made its appearance, but probably the geological equivalent of the red and variegated sand rock. It is a very coarse but stratified conglomerate, with pebbles of agate, quartz, trap, amygdaloid trap, red granite, &c., many of them larger than a man could lift. It is raised in *uplifts* corresponding with the subordinate trap, and contains fissures like the trap, which are filled with spar. The general course of the uplifts are south-west by west, and the course of the fissures or veins, both of the trap and conglomerate, is nearly at *right angles* to the face of the uplifts. It is in these veins that the native copper and its ores are found.

The line of greatest elevation runs near the middle of the point, forming an anticlinal axis, from which the rocks pitch each way, at various angles, from twenty to sixty degrees. But it must not be supposed that the descent is regular from the summit towards the lake. In the volcanic convulsions that generated and raised the trap rocks, they were greatly broken and fractured, and consequently the overlying rocks, the conglomerate and sandstones, were dislocated in the same way. They now lie in the form of vast steps, the broken faces of the conglomerate and trap nearly perpendicular, and the slopes at the angles above stated. The veins of the stratified and the unstratified rocks appear to be of the same age ; to have been formed by the same cause, after the enclosing rocks had taken the form and position they now have. Upon the manner of the formation of these veins there are various conjectures, which I have not space to notice. When they pass from the conglomerate to the harder and more compact trap rock, they are said to diminish in width, and the material of the vein changes. They carry, in general, beautiful calcareous spar, and also other substances besides copper, such as quartz and barytes.

From the Manitou Islands, at the extremity of Point Kewena, to the Portage Lakes, the most elevated mountain range, or rather succession of knobs, is nearer to the north than the south shore, and from 700 to 800 feet in height. It is a very rough region to explore, with precipitous rocks, thick cedar swamps, and tangled evergreens in every part. But Dr. Houghton, with five companies of explorers and surveyors, has subdivided all the land east of the Portage Lakes into sections during the past summer, except one fractional township. The labor and exposure attending this work cannot be understood by any except those who have been upon the ground and seen its mountains and swamps. This survey was undertaken to demonstrate the practicability and value of a favorite system of Dr. Houghton's. He had, as geologist of the State of Michigan, spent several years in this desert region, and knew its mineral worth. He felt, as every exploring geologist feels, the necessity of exact topographical and lineal surveys, in order to give his reports that character of perfect accuracy of which the science is capable. In truth, a large portion of the results of mineral explorations is geographical, topographical, and mathematical matter. The thickness, extent, and dip of rocks, when found, constitute a perfect measurement of the country. Dr. Houghton contracted with the Government to make the *lineal survey* of this region, and at the same time a *geological* one; and laboring upon it as the great undertaking of his life, had, as I have remarked, nearly completed the most difficult portion — that of Point Kewena. His melancholy fate is well known.

By these surveys, Fort Wilkins and Copper Harbor are situated near the south-west corner of town 59 *north*, range 28 *west*, or twelve towns north, and twenty-eight west of St. Mary's.

The returns of the Government surveys of this region will not only show the coast and water-courses in connection with town and section lines, but will give the elevation and depression (what public surveys hitherto have not) of the country, taken at every change, by the barometer. They will further exhibit the exact limit and character of the mineral region. Such a system introduced into all the public surveys, with modifications suitable to the agricultural districts, such as the analysis of soils, collection of plants and marls, would be of immense advantage to the settler, and honorable to the nation.

The maps and papers of the mineral agency at

Porter's Island, in Copper Harbor, showed about 500 locations of one mile square each. The War Department has, by usage, the control of the mineral lands of the United States. It is doubtful whether there is any law that covers the case of the copper mines of Lake Superior. The President has, however, reposed the power of leasing these and other mineral lands in the War Department, which confines their management to the Bureau of Ordnance, which acts by local agents. The Secretary of War, or the local agents, grant permits of search and location, and the location being made, a lease is granted to the locator. In this lease there are covenants to render the Government *six* per cent. of the mineral raised for three years, and after that time the Government have power to require *ten* per cent. for the next *six* years.

At first the permits included three miles square, or nine square miles, but were early last spring reduced to one square mile, and given upon every application, without fees. About seventy permits were now laid in the neighborhood of Dead Men's River, and eight or ten about the mouth of Huron River. The Point Kewena proper, that is to say, that portion east of the Portage Lakes, was mostly covered, and various other large tracts on the waters of Elm River, the Ontonagon, Iron River, Montreal, and even on the Brule, beyond La Pointe.

In order therefore to locate our permits, it became necessary to go westward and explore some of the vacant regions beyond the Portage Lakes. We therefore left Copper Harbor, touching at Agate Harbor, Eagle Harbor, and Eagle River, and proceeded to the mouth of Salmon Trout River, in township 55 north, range 35 west.

Mr. Bushnell and myself, and two men, here took to the woods, and striking the range line between 34 and 35, followed it south to the south-east corner of township 53 north, range 35 west, being about seventeen miles interior. To our surprise instead of finding a rugged mineral region, we had passed through a handsome rolling country, tolerably well watered, with a good loamy soil, producing an abundance of sugar-maple. Along the margin of the lake, owing probably to the harsh and moist winds from the water, nothing but birch, balsam, pine, hemlock, spruce, and white cedar is seen; but at the distance of two to five miles interior the forest growth changes entirely. There is an occasional white pine, with a lofty, straight and majestic trunk, some scattering elms, linns, and black oaks, but the reigning tree is the sugar-maple.

On our left lay the valley of the Portage Lakes and of Sturgeon River, which we had just crossed. Turning westward we soon encountered one of those eye-sores to the explorer and surveyor, a *cedar swamp*, in which a progress of a mile an hour is considered rapid travelling. The white cedar lives to a great age before it begins to decline. It finally rots at the root, and is blown down by the northern tempest. But this is by no means its end; its prostrate trunk sends up live branches, that draw sustenance through the roots of the parent, or new prongs sent by itself below, among the buried trunks of preceding centuries. In after ages, when it has at length matured, and weakened by time has yielded to the winds, another sprout from its side keeps the family stock in perpetual being. Beneath the accumulated bodies of these trees, some dead and some living, the water, in which they delight, stands the year through, flowing gradually towards some stream of the vicinity. What is remarkable, the water of these swamps, so little and slow is the decay of the cedar tree, is clear, pure, and cool.

I hope I have been able to convey to the reader

a just idea of a white cedar swamp, because without a correct conception of this, he will never be able to realize the great difficulty of travelling in this new country. After he has penetrated one of them forty rods, the view is equally extensive in every direction, whether it is only forty rods to the other side, or whether it is two miles. In addition to the net-work of logs, and the thicket of leaves that never fall, it is necessary to think of numberless dry, sharp, and stiff prongs, the imperishable arms and limbs of dead and fallen trees. It is then to be remembered that every man carries more or less of a load upon his back; his blanket, his tin cup, probably some implement, a hatchet or a hammer, with specimens, and a few pounds of provisions.

The second night found us advanced about one mile into a noble cedar swamp. Climbing a tree extended somewhat the range of the eye, but it met only the sombre and half naked trunks of the white cedar, in every direction. A camp-bed was formed beneath a tall and beautiful *larch*, or tamarack, and a fire made at its root. The bed was made as usual of branches, kept out of the water in this instance by brush and poles. This white cedar has the merit of burning readily, as well as of durability, and made to-night a bright fire, flaming gaily upwards against the straight and stately larch. When had such an illumination shone there before? The owl gave utterance to his surprise in hideous screams, and hooted for his mate. The larch, as it swayed to and fro in the night breeze, seemed to creak and groan because of the fire, which was scorching its sinews and boiling its life-blood in its veins. No doubt before many seasons pass by he will sicken and die, and from a tall prince overlooking the humble cedars, will come heavily down, perhaps in the stillness of night, and lay his body along side of theirs.

In the morning, after passing a cold and comfortless night, a few minutes travel cleared the swamp, and rising some very high land we found the stratified sandstone again, and inclined towards the lake.

At the south-east coner of township 53 north, range 36 west, the trap-ranges again made their appearance, from whose summits the mountains of the Huron River were visible in the south beyond the Anse.

We were now on the head waters of Elm River, on ground located for many miles around. Most of them are what are called office locations, made without visiting the spot, and in consequence of some locations made by Mr. Kenzie, of Chicago, from actual observation, of which favorable reports were in circulation.

That night we should have met two of our men at a rendezvous with supplies; but neither party had sought the right spot, so indefinite were the descriptions given us of localities. As it was some miles from the coast to the mineral ranges, the boat passed slowly along the shore, sending out provisions from time to time to the exploring party. It was not then known how far west the township lines were surveyed, consequently the points of meeting were fixed at the forks of some stream, or some old camp, and in finding these many errors might be committed. In this case a day was consumed in uniting the two parties, which would not have been of so much consequence, had not the stock of eatables began to fail. But most of the disagreeable effects of a short allowance were avoided by the capture of a porcupine, of which we made by long boiling in the camp-kettle, a very palatable soup.

On the 20th of September, at a distance of twenty miles from the coast, there were a few flakes of snow, succeeding a cold rain. On the 21st and 22d, rain. The ground passed over during this week is drained by the Salmon Trout River (a creek), Elm River, Misery River, Sturgeon and Flint Steel Rivers. Every member of the party was delighted with its soil, its beautiful and heavy timber, and the unsurpassed purity, plenty, and coldness of its waters. We passed several small clear lakes, the sources of many streams. These streams are in general but few miles in length, enlarging very fast as you follow them downward from the head, alive with the famous speckled trout, rapid in their descent, and so uniform in the flow of water, that water power is everywhere abundant. Many a time did Patrick and Charley select their future farms on the border of some quiet pool, from which a tumbling brook issued, bearing its faithful tribute into the reservoir of the Father of Lakes.

The cedar swamps so hateful to the explorer will be necessary to the farmer for his supply of rails; the tall round pines scattered here and there among the sugar trees, now so green and majestic, will supply him with lumber; the straight and beautiful balsam with timber.

Hitherto the mineral trap rocks that rise occasionally through the sandstone stratum, do not greatly interfere with the use of the land for tillage. This rock, when fully disintegrated, gives a light soil that produces well. In this vicinity the trap rises suddenly out of the plain land, sometimes with one perpendicular face and one gentle slope, sometimes like an island with a bluff all around, and flat, rich land on the top, and sometimes in irregular peaks, standing among the timber like cones and pyramids. At the sources of Flint Steel River we saw, interspersed with protruding summits of trap, peaks of conglomerate, shooting up from flat land to the height of fifty, seventy, and one hundred feet.

Pursuing a south-westerly course, about noon on the 26th we entered the ravines that lead into the Ontonagon. From Elm River to the Ontonagon, the sand rock is covered from ten to four hundred feet in depth with a stratified deposite of *red clay and sand*, very fine. It is commonly called clay, but contains more silex than alumine, though it is so minutely divided as to have the appearance of clay. I saw nowhere true clay beds, but it is possible some of this deposite will harden in the fire so as to make bricks.

This great sand-bed is easily washed out by running water. From the Falls, the Ontonagon has hollowed out for itself a channel 300 to 400 feet deep, and from half a mile to two miles wide. The lateral gullies are very numerous, deep, and steep. Every permanent rill, operating for ages, has excavated a narrow trough, the bottom of which descends towards the river, in the inverse proportion to its length, and the sides remain as nearly a perpendicular as the earth will lie. The low grounds not so wet as to cause cypress and cedar swamps, are everywhere inclined to produce hemlock and balsam. It is the same in the ravines, cold, moisture, and a confined atmosphere, causing the growth of evergreens, and also of cedars.

It will now be easy to judge of the facilities of travelling in the region of the gullies. To cross them, rising one slippery face and sliding down the next, is very exhausting to men loaded with packs. To follow down one of the ravines, so narrow, deep, and shaded, as almost to exclude the sun at noon, is much like the change "from the frying-pan into the fire." The timber of the sides has fallen inward, into and across the contracted pathway of the rivulet, so thick and so much entangled, that the mind is in a constant state of exercise, determining whether it is easier to crawl under, or climb over the next log.

In such regions, as you approach the common discharge of all these ravines, as a creek, a lake, or, as in this case, a river, the number of lateral gullies diminish, and it is sometimes preferable to take the crest of the gulf and follow it towards the mouth. We did so, and coming along a narrow backbone, scarcely wide enough for two to walk abreast, suddenly came to its termination, with the river far below us. It was noon of a lovely day, such as are called the Indian summer. In the distance to the north, twelve or fifteen miles, a thick haze covered the lake; the sides and bottom of the valley of the Ontonagon were brilliant in the mellow sunlight, mottled with yellow and green; the golden tops of the sugar tree mingled with the dark summits of the pine and the balsam. The rough gorges that enter the valley on both sides were now concealed by the dense foliage of the trees, partly gorgeous and partly sombre, made yet richer by the contrast, so that the surface of the wood as seen from our elevation, in fact from the waving top of a trim balsam which I had ascended, lay like a beautifully worked and colored carpet ready for our feet.

On this promontory, jutting into the valley, we kindled a fire in the dry and hollow trunk of a hemlock, as a beacon to our companions, who were to be at the foot of the rapids with the boat.

On the left or inland side the valley at some miles distant is seen to divide, corresponding with the two branches of the river. In this direction are elevated peaks, several hundred feet higher than our position, but partly hid in the mist of the atmosphere. We had now spent as much time in scene-gazing as was profitable, and taking up our packs tumbled down the bluff to the river. There stood the tents, and there lay the boat, with our comrades lounging about in the sun. The meeting brought forth three hearty shouts all around, and such congratulations of genuine good-will, as none but woodsmen and sailors know.

We were now at the foot of the rapids, one mile north of the correction base, which is also the line between towns 50 and 51 north, and one mile east of the range line between ranges 39 and 40 west.

On the next day, after washing, drying, and mending some of the most needed garments, Patrick, our faithful Irishman, and myself, crossed the river and went west along the correction line. This course carried us constantly nearer the lake, because the direction of the shore is south of west. The timber was, as might have been expected on approaching the lake, more hemlock, birch, and balsam, but the soil appeared as good as that we had passed over from Salmon Trout River in range 35 west. In range 41 west, we turned to the left, and soon found that no surveys had been made south of the correction line. The same day a rain set in that lasted with little intermission four days and five nights. In the trap region, the magnetic needle is subject to great fluctuations. When the sky is overcast, as it was in this case from morning to night, the sun, the principal guide, is of course lost. If the traveller loses his *confidence* in the compass, that instrument is the same as lost, and he is compelled to rely upon judgment, or rather the woodsman's instinct. This judgment is sometimes a very uncertain reliance. The streams and ridges of land are so irregular that little information can be drawn from them.

There is a great difference in persons in the accuracy of their calculations, guided by the "make of the country," as its general topography is called.

In this region, none but the oldest hunters and trappers feel safe when the compass begins to play false and the sun withdraws himself. If the consumption of provisions could cease for the time, it would always be safer and wiser to stop and encamp until clear weather comes, but the appetite does not seem to know that circumstances alter cases. With the mind in a state of perplexity, the fatigue of travelling is greater than usual, and excessive fatigue in turn weakens not only the power of exertion, but of resolution also. The wanderer is finally overtaken with an indescribable sensation—one that must be experienced to be understood—that of *lostness*. At a moment when all his faculties, instincts, and perceptions are in full demand, he finds them all confused, irregular, and weak. When every physical power is required to carry him forward, his limbs seem to be yielding to the disorder of his mind; he is filled with an impressive sense of his inefficiency, with an indefinite idea of alarm, apprehension, and dismay; he reasons, but trusts to no conclusion; he decides upon the preponderance of reason and fact as he supposes, and is sure to decide wrong. If he stumbles into a trail he has passed before, or even passed within a few hours, he does not recognize it, or if he should at last, and conclude to follow it, a fatal lunacy impels him to take the wrong end. His own tracks are the prints of the feet of some other man, and if the sun should at last penetrate the fogs and clouds that envelop his path, the world seems for a time to be turned end for end; the sun is out of place—perhaps it is to his addled brain, far in the north, coursing around to the south, or in the west moving towards the east. At length, like a dream, the delusion wears away, objects put on their natural dress, the sun takes up its usual track, streams run towards their mouths, the compass points to the northward; dejection and weakness give place to confidence and elasticity of mind.

I have twice experienced what I have here attempted to describe. It is a species of delirium. It oppresses and injures every faculty, like any other intense and overwhelming action. The greatest possible care should be taken to prevent the occasion for its return. Two men, last summer, were exploring on Elm River, and without compass or food started for a vein a few rods from camp. They got entangled among swamps and hills, and wandered forty-eight hours in the woods, bewildered and lost. By accident they struck the lake shore, and their senses returned. It is not prudent to be a moment without the means of striking a fire, without food for a day or two, and a plenty of clothing, or without a compass. Martin and myself went out in the morning from Salmon Trout River, intending to go three miles and return. He had neither coat, nor vest, nor stockings, because the weather was mild. A rain soon came on, and a thick mist; steering for the camp, we struck the creek two miles above the mouth and the camp. The ground in the vicinity of the lake has a low evergreen bush, with a leaf like the hemlock, which lies flat on the surface, entangling the feet at every step. It was dark when we struck the creek, and began to follow it down stream. The sloughs, logs, ground hemlock, and cedar brush were so bad, that it would have been difficult to make much progress in daylight, and it was now pitch dark. We took to the water-course to avoid the brush and bluffs of either bank, and waded along the channel. But the waters of these streams are always cold, and Martin, though a stout fellow, and full of resolution, began to be numb with cold and wet. We had nothing to eat, our matches were wet, the gun could not be fired off. There was but one course to pursue. The stream would take us to camp, but how far distant that desirable spot lay we could not conjecture. But the chilly water must be avoided, and the brush and logs, wet, slippery, and numberless as they were, must be surmounted. "We have crossed that log before," says Martin. "*What, are we lost?* impossible; we have not left the stream a moment—it cannot be." Crooked and winding as it was, it is not possible that we should travel twice over the same ground. But there was the log, to all appearance the same we had crossed half an hour before. Both of us would swear to the identity of the log—the same timber, the same size, the same splinters at the root, the bark off in the same way; and still it was more probable that *two* such logs should be found, than that we had passed *twice* over the same spot.

We crawled onward, filled with the mystery—and it is not to this hour anything else than a mystery. In about two hours my companion gave an exclamation of hope and joy. He had been up the creek the day before shooting ducks and fishing for trout. He recognized the spot where the canoe was obstructed by flood-wood, half a mile from the tents. We now knew where there was a trail, and in a few minutes beheld the sparks of the camp-fire ascending gaily among the trees.

With fire works better secured, with more attention to clothing on the part of Martin, and to blankets by both of us, especially with ordinary prudence in regard to provisions, the discomfort and exertion, the bruises, chills, and exhaustion of this day, so injurious to the constitution, whether felt immediately or not, might have been entirely avoided. It may be thought that such vexations might be prevented by a rational foresight, and this is no doubt true; but in practice they occur frequently to woodsmen, and they are in general as keen in the examination of chances as any class of men. Even Indians and Indian guides become bewildered, miscalculate their position, make false reckoning of distances, lose courage, and abandon themselves to despair and to tears.

The maps for the copper region, instead of assisting the explorer, were for the interior so erroneous — a fault worse than deficiency — that mistakes equal to a day's travel frequently resulted from a reliance upon them.

On the office map there was noted a lake, not far above the forks of the Ontonagon, on the west fork. Leaving the "correction base" at the south-west corner of town 51 N., range 40 W., we should have struck that lake in the distance of ten miles; but instead of a lake, found ourselves involved in the marshes at the sources of the Cranberry and Iron Rivers, the lake itself being about fifteen miles distant. The forks of the Ontonagon appeared from the map, and the best information within reach, to be about four miles by river above the foot of the rapids. This was made a point in our return, to which a packer was sent with pork and beans. Instead of making the rendezvous in one day's travel, as was expected, he reports the distance at fifteen miles by river, and seven or eight in a direct line. The delay occasioned by bad weather and mistakes amounted on our part to two days; the packer, who had at last reached the forks, after spending two nights in a cold rain, without fire, had left, and carried back his provisions. Patrick had, by mistake, taken salt pork for three men instead of two. When we arrived at the Forks, only one meal of bread and beans remained, with a little tea and sugar; but the pork was sufficient for two days more. It was necessary to alter our route, and employ those two days in reaching the agency at the mouth of the river. This is an instance of hazard and disappointment, and it is difficult to see how it could have been avoided. With the greatest sagacity and forethought, small parties, who do not survey and mark their courses and distances, cannot avoid occasional perils.

The circumstances in which we were placed did not allow of as much observation upon that interesting region, the Falls of the Ontonagon, as I desired. The greatest fall is on the west branch, and occupies a distance of at least two miles, with a descent of about eighty feet. It was at the head of this succession of cataracts, that the "Copper Rock" was found, which is now at Washington City. It lay, when first discovered, on the brink of the river, in the red clay deposite, of which I have spoken, although mountains of trap, sandstone, and conglomerate rise on all sides. The rock was removed from its place upon a temporary railway, constructed through the woods about four miles, to a point on the river where it could be floated. This road crossed deep ravines, and a steep mountain 300 feet high. The rock was hauled along on a car, and up the mountain by a capstan and ropes. Its weight is a little over 3,000 pounds.

It is now eighty years since this copper rock obtained notoriety among white men. Mr. *Alexander Henry*, an adventurous Englishman and an agreeable writer, who entered the Indian country immediately after the peace of 1763, gives a description of the rock which is worthy of being repeated.

"On the 19th of August (1765) we reached the mouth of the River Ontonagon, one of the largest on the south side of the lake. At the mouth was an Indian village, and at three leagues above a fall, at the foot of which sturgeon were at this season so abundant, that a month's subsistence for a regiment could have been taken in a few hours. But I found this river chiefly remarkable for the abundance of virgin copper, which is on its banks and in its neighborhood, and of which the reputation is at present (1809) more generally spread than it was at the time of this, my first visit. The copper presented itself to the eye in masses of various weight. The Indians showed me one of twenty pounds. They were used to manufacture this metal into spoons and bracelets for themselves. In the perfect state in which they found it, it required nothing but to be beat into shape. The 'Pi-wa-bie,' or Iron River, enters the lake to the westward of the Ontonagon, and here it is pretended silver was found, while the country was in the possession of the French."—Part I, pp. 194–5.

"On my way (1766) I encamped a second time at the mouth of the Ontonagon, and now took the opportunity of going ten miles up the river with Indian guides. The object which I went most expressly to see, and to which I had the satisfaction of being led, was a mass of copper, of the weight, according to my estimate, of no less than five tons. Such was its pure and malleable state, that with an axe I was able to cut off a portion weighing a hundred pounds. On viewing the surrounding surface, I conjectured that the mass, at some period or other, had rolled from the side of a lofty hill which rises at its back."—p. 205.

I quote extensively from Mr. Henry's interesting book, because it is now out of print, and very rare. Captain *Jonathan Carver*, also, travelled in the Lake Superior and Mississippi country in 1766, of whom, after the manner of succeeding travellers, speaking of their predecessors, Mr. *Henry* says, "and he falls into other errors." The Chippeways told Carver, that being once driven by a storm to the *Isle de Mau-*

repas (now *Michipicoten*), they had found large quantities of shining earth, "which must have been gold dust." They put some of it into their canoes, but had not moved far from the land, when a spirit sixty feet in height strode into the water and ordered them to bring every particle of it back to the island. This of course they did, and never ventured again to the haunted island.

In the spring of 1769, Mr. Henry, excited by this and other reports of the Indians, visited the islands, expecting to find "shining rocks and stones of rare description," but found only a mass of rock rising into barren mountains, with veins of spar. The Indians then insisted upon going to another island to the south (Caribeou) as it was the true island of the "golden sands;" but the weather prevented this visit at that time. In 1770, Mr. Baxter, Mr. Bostwick, and Mr. Henry, were constituted members of a company for working mines on Lake Superior.

"We passed the winter together at Sault de Sainte Marie, and built a barge fit for the navigation of the lake, at the same time laying the keel of a sloop of forty tons. Early in May, 1771, we departed from Point aux Pins, our shipyard, and sailed for the island of Yellow Sands, promising ourselves to make our fortunes in defiance of the serpents. I was the first to land, carrying with me my loaded gun, resolved to meet with courage the guardians of the gold.

"A stay of three days did not enable us to find gold, or even yellow sands, and no serpents appeared to terrify us, not even the smallest and most harmless snake.

"On the fourth day, after drying our Caribeou meat, we sailed for *Nanibojou* (on the north shore), which we reached in eighteen hours with a fair breeze. On the next day the miners examined the coast of

Nanibojou, and found several veins of copper and lead, and after this returned to Point aux Pins, where we erected an air furnace. The assayer made a report on the ores which we had collected, stating that the lead ore contained silver in the proportion of forty ounces to the ton; but the copper ore only in very small proportion indeed."

The party now start for the Ontonagon, having in company a Mr. Norberg, an officer in the 60th regiment then stationed at Mackinaw, old fort. At Point Iroquois he found among the loose stones one "of eight pounds, of a blue color, and semi-transparent," which he deposited in the British Museum at London, and which, it is said, contained *sixty* per cent of silver.

"Hence we coasted westward, but found nothing till we reached the Ontonagon, where, besides the detached masses of copper formerly mentioned, we saw much of the same metal imbedded in stone. Proposing to ourselves to make a trial on the hill till we were better able to go to work on the solid rock, we built a house and sent to the Sault de Sainte Marie for provisions. At the spot pitched upon for the commencement of our preparations, a green colored water, which tinged iron of a copper color, issued from the hill, and this the miners called a *leader*. In digging they found frequent masses of copper, some of which were of three pounds weight. Having arranged everything for the accommodation of the miners during the winter, we returned to the Sault. Early in the spring of 1772, we sent a boat-load of provisions, but it came back on the 20th day of June, bringing with it, to our surprise, the whole establishment of miners. They reported that in the course of the winter they had penetrated forty feet into the hill, but that on the arrival of the thaw, the clay on which, on account of its stiffness, they had relied, and neglected to secure by supports, had fallen in; that from the detached masses of metal which to the last had daily presented themselves, they supposed there might be ultimately reached some body of the same, but could form no conjecture of its distance. Here our operations in this quarter ended. It was never for the exportation of copper that our company was formed, but always with a view to the silver which, it was hoped the ores, whether of copper or lead, might in sufficient quantity contain."—pp. 227, 232.

"In the following August we launched our sloop, and carried the miners to the vein of copper ore on the north side of the lake (probably at Nanibojou, about one day's sail from Michipicoten). Little was done during the winter, but by dint of labor performed between the commencement of the spring of 1773, and the ensuing month of September, they penetrated thirty feet into the solid rock. The rock was blasted with great difficulty, and the vein which at the beginning was of the breadth of *four* feet, had in the progress contracted into *four* inches. Under these circumstances we desisted, and carried the miners back to the Sault. What copper ore we had collected we took to England, but the next season we were informed that the partners there declined entering into further expenses. In the interim we had carried the miners along the north shore, as far as the river *Pic*, making, however, no discovery of importance. This year, therefore (1774), Mr. Baxter disposed of the sloop and other effects of the company, and paid its debts. The partners in England were his Royal Highness the Duke of Gloucester, Mr. Secretary Townshend, Sir Samuel Tucket, Baronet, Mr. Baxter, Consul of the Empress of Russia, and Mr. Cruikshank. In America, Sir William Johnson, Barronet, Mr. Bostwick, Mr. Baxter, and myself. A charter had been petitioned for and obtained, but owing to our ill success, it was never taken from the seal office."—pp. 234–5.

There is living an old chief who, when a boy, saw this company of English miners at the falls of the Ontonagon. He represents the manager as a stout burly man with a red face. There are near the spot where the great copper rock was found, remains of a chimney, supposed to belong to the house spoken of by Henry. The timber around the spot was of a second growth, now cut away by Mr. James Paul, who has lived there and located a three-mile permit. He told me that an aspen, eighteen inches in diameter, had blown down near his cabin, and a copper kettle was found flattened and corroded beneath its roots. There are also the remains of ancient pits still visible, and in the sand and clay deposite, by digging, lumps of native copper are now found. There can, therefore, be no doubt but this is the spot visited by the English company before the American Revolution, and now become again an object of hope and notoriety.

This region is singularly wild and disordered. The Falls, which are distinct from the "Rapids," are caused by the irregular upheaval of trap, sandstone, and conglomerate, thrown about in grand confusion. To the miner and geologist such points possess not only the greatest interest, but the greatest practical value.

Here appears to be one of those great centres of convulsion which raised and tossed about the metalliferous rocks. Another may be seen to the eastward of the Portage Lakes. From the central point in each direction along the line of action, that is to say, in a north-easterly and south-westerly course, the height of the upheaval and the extent of the distortion gradually becomes less on each side. The effect of the subterranean forces being very much the same upon the overlying sand-rock, as that of a projecting point of rock upon the ice of an estuary of the sea when the tide falls away. The trap-uplifts represent the rock, itself rising instead of the sandstone stratum settling. The resemblance is not perfect, but only illustrative. The field of ice subsiding upon a sharp point of rock, in a bay of quiet waters, will break and crack equally in all directions; but the uprising trap, though it has a centre, does not act equally on all sides; for there is a *line* of upheaval, along which the force operates, giving rise to an elevated ridge, which is highest at the centre, or focus. It has a breadth of five to fifteen miles, and a length of fifty or sixty. The trap-rock intruding from below, has within itself a certain regularity, which I have noticed before: throwing up long parallel faces, looking inward towards the line of greatest elevation.

On this fact, I have from observation a knowledge of only a portion of the northern half of the trap range, from the Manitou Islands to Iron river, a distance of about 120 miles. I did not cross the range far enough to ascertain the position of the southern half, and give this statement of its organization upon the representation of other explorers, whom I have no reason to doubt.

These ranges are not in every case parallel to the great anticlinal line, but generally they are so. There are cases of spurs, or lateral ranges, of limited extent, branching off from the main pile. Both the trap and the overlying conglomerate rocks are very hard to work. The trap is the most compact, but is more uniform in its texture. The conglomerate encloses pebbles of all sizes, and of many different rocks, most of them very hard. This want of homogenity prevents the blast from producing that effect which it would on a close, uniform, tight rock. I think there can be little doubt but Mr. Henry's conjecture respecting the source of the copper rock of the Ontonagon, and the many copper boulders, found in the red clay deposite, is correct. That they were loosened from their position in a neighboring vein, by the disentegration of the enclosing rock, and by the force of gravity and that agent, whatever it may have been, which brought on the red sand and clay deposite, they have been scattered around. The red deposite is evidently younger than the sandstone and the trap, for it is horizontal. The sandstone, it is equally evident, is *older* than the trap, for the latter has shot up through it, tilting it outward from the line of uplift. The copper boulders are found imbedded in the red loam, as it may be called, and must have been loosened from the vein at and before the period, when it (the loam) was brought on.

The native copper, which is the principal ore of the country, (if metal can be called an ore,) exists in the veins in all sizes and shapes, from the weight of the point of a pin to 20, 40, 100, 1,000 and 1,500 pounds. A boulder was found this season, near the mouth of Elm river, weighing over 1,500 pounds, which is now at New Haven. I saw an irregular mass in a vein near Agate Harbor, about one mile east, which might, with great care, have been taken out, weighing 800 to 1,000 pounds. It was removed in one body, to the amount of 400 pounds; but to procure such specimens, there is great trouble and expense in securing all the prongs against damage by the blast. These boulders are found in the water-worn pebbles of the shore, and of various sizes, from one to forty and one hundred pounds. They are also found far to the southward, in Wisconsin—giving rise to great hopes and speculations—transported by that universal power (whatever it was) which covered the northern hemisphere with drift from the north.

It may then be suggested, whether the great copper rock and its satellites of the Falls of the Ontonagon were not carried thither in the same manner. There is certainly room for such a doubt. But no matter how far these masses of copper have been transported, or how short the distance they have been moved, they must have originally been derived from veins. Here we find not the particular veins from which the boulder was extracted, but find in the country veins containing exactly such masses. They may have been dragged from regions farther north, where similar veins probably exist, but as there is no necessity for going to so great a distance in search of their origin, so there is not as great a probability of finding their original seat far from their present position. The difficulty of transporting such heavy material is a strong reason against distance, though not a conclusive one.

But in the case of the great rock, the number of attending fragments is so *numerous*—so much more so than is known anywhere else at a distance from the veins, that little doubt remains that they are from a nest not very far off. In the gold region, and in the lead mines, where loose metal is found, the miner begins to search in all directions to ascertain from whence it came. If he finds it more abundant on one side than another, he examines more closely the soil of that side; and if found to increase as he proceeds, he is convinced that he is on the trail. As he follows this, the evidences multiply, and at last he arrives at the parent vein, from which the scattered fragments were driven. It is probable that time, money, and enterprise, will finish what the English company began, and at last disclose a prominent vein within hearing of the cataracts of the west branch.

The mouth of the Ontonagon is one of those commanding points that strike the observer at first glance. As Henry says, it is the principal river of the south shore, and the only one, except the Chocolate river and Grand Marais, where a vessel can enter. There is now, in a low stage of the lake, six feet water on the bar, and deep water several miles up the stream, which is about 300 feet wide. It is the natural outlet of a large farming region, which the surveyors say, extends fifty or sixty miles interior, and forty or fifty each way along the shore. The mineral belt occupies several miles in width, at this point ten or twelve miles from the shore, and parallel with it; but at the mouth of Iron, Black, and Montreal rivers, it comes down to the waters of the lake. On each side of this range, and even among the Porcupine Mountains, the agricultural resources of the country are only limited by the shortness of the seasons. The soil is good—the climate without an equal for health and strength, and the lake and streams abound in fish. The swamps and the flat lands produce wild grass in abundance, showing the tendency of the soil to that production. Potatoes, turnips, and all roots grow here in the greatest perfection, and oats and barley do well. I have little doubt but it will also be found an excellent wheat region.

We found the rich bottom-lands of the Ontonagon already dotted with the cabins of pre-emption claimants, for several miles up the river. The Indians have a tradition about the name of Ontonagon, as about almost every thing else, and say it is truly "Nindinagon." That an old woman, long ago, was cooking on the shore at the mouth, and her dish slipped into the current and was carried out into the lake.

She exclaimed, "Oh! there goes my dish," the Indian of which is said to be Nindinagon.

The site at its mouth is rather low and swampy. On the west the Porcupine Mountains rise boldly out of the water, at the distance of twenty miles, presenting that peculiar outline of the trap uplifts, by which they may be recognised afar off almost as well as by inspection. A cross-section, which would also correspond with the end view from the Ontonagon, may be compared to the notches or teeth of a mill-saw laid upon its back, one edge straight and vertical, the other sloping. If the expectations of mineral locators are realized, the prosecution of the mining business will of itself create a place of some importance here. To the farmer of New England there will be great inducements, as soon as the mining operations are placed upon a sure footing; for the products most congenial to the region are such as are bulky, and cost much in their transportation, to wit: potatoes and roots, hay and oats. It is well known that miners never till the soil to much purpose. A garden and a little pasture suffice for them. This must be done by the practical farmer. The mineral and the agricultural districts are here so admirably situated, as mutually to render to each interest the greatest assistance. When the navigation shall be completed around the rapids of the St. Mary's, the emigrant and miner, placing himself at any harbor of any of the lakes, may take his passage to any part of Lake Superior, with his family and effects. The hardy son of Vermont and New Hampshire will find here his own climate and mountains, his own trout streams, and a good substitute for the shad and salmon of the ocean; and a soil, equal to most parts of the West, without the fever and ague of the more southern portions. The facility of making roads to the interior is great, and along the shore they are practicable. Of course, on the immediate coast, ravines are too frequent to cross without expensive bridges. But a few miles inland, the country rises, the valleys of the streams diminish, and a very favorable country is found as far east as the Portage Lakes and the Anse. Here the swamps and lakes form the only serious obstacles, and they are avoided by good selections of routes. The difficulty of making roads in the Ontonagon region is far less than it was in the first settlement of Ohio.

Until the night of the 5th of October I had not observed any frost, although the leaves were already colored with the hues of autumn, and falling from their stems, had begun to cover the ground. The winds and rains that occurred between the 5th and the 10th left the branches of the trees almost as naked as in winter, and the snow began to fall. We were received at the Agency house with that liberality of hospitality which can be found nowhere more full and hearty than among the backwoodsmen of the West. Major Campbell, the agent, was absent in search of a copper rock, in the neighborhood of "Lake Vieux Desert," about 150 miles distant. In the evening, Mr. Paul, who has been three years in the country, and who had joined in the wild-goose chase after the copper rock, on the faith of an Indian, came in, and amused the company till a late hour by reciting the stratagems and effrontery of their Indian guide.

Since the whites have shown such an intense curiosity about copper rocks, they have sprung up on all sides. Every Indian knows where one may be found. It can be had of any size or shape, and generally for the price of a few dollars and provisions for the trip. It is generally seven, ten, or twelve days' journey to it. The Great Spirit and the tribe will destroy or otherwise injure him who shows it to the white man, but they will lead him to the vicinity, and he can do the rest. In this case a monster was to be found, and the price was to correspond; but fifty or sixty dollars was somehow procured in advance. The Indian lived in the neighborhood of the rock and had shown it to but one other mortal, a half-breed, now dead. After great labor and vexation, the party approached the sacred place. There are four trees marked with porcupines, done in charcoal, according to the description. They were far from any trap ranges, in a low, swampy country. The Indian fixes his eyes in a given direction, and all are elated with a certainty of success. They scour the woods in that direction, but no rock is found. The Indian and his boy wish to be left to pursue the search by themselves, and still the rock hides itself. He is watched, and they find that he only moves around in a limited circle, and returns to the camp. Hesitating between the apprehension that he is duped, and the realization of his hopes, the agent becomes impatient. The Indian at length points his finger to the spot, but the Great Spirit had sunk the rock deep into the earth. The Indian is calm and immovable. "*Hou, hou—march on wigwam,*" he says, in the usual tone. "What does he say?" inquires the agent. "He says, we had better go to his wigwam," replies the interpreter. The scene changes from the highest expectations to the highest rage. "Give him a hundred lashes—break every bone in his body—*kill him!*" and expressions of this sort are now heard, with gestures to match. The Indian could not understand English, but knew enough to be sensible that some cursing was going on, and that he was the object. He now began to kindle with wrath. The first motion was to throw down his pack, and in this he was followed by the boy and two or three other Indians of the party. What was the agent, the surveyor, and the interpreter to do, here in this wilderness, deserted by their packers and guides. Paul, who had long known the Indian's cunning, saw at once the position of affairs, laughed at the agent, and offered the Indians a half dollar to take up their packs. They had, in the mean time, proceeded from anger to mockery. They had paraded themselves in advance of the party, strutting along with some small willow sticks on their shoulders, in derision of the heavy loads under which the whites were groaning. The latter were obliged not only to pocket the insult, but to employ the old man, his boy, wife, and canoe, to cross some lakes that lay in their route home.

Coming in they met another party of whites, with the usual complement of Indians, also in search of a copper rock, said to exist in the region of *Lake Vieux Desert.* If such rocks were actually visible, no Indian would show it, so long as he can get one-half of his yearly support from it as a guide. Those who know them best, say that it matters little to the explorer whether such boulders exist or not, the Indians will never be guilty of showing one to a white man. There is a superstition upon the subject, and it is also a rule that the proceeds of a found rock should be divided, and a large portion go to the chief. In case an Indian actually knew of one, he would not disclose its position, unless he was sure the fact would never be made known to his tribe.

On the morning of the second day the square-sail of our boat, which had been to La Pointe, appeared at the foot of the Porcupine Mountains, bright in the light of the rising sun. At eleven it entered the river, before a bountiful breeze, and the company was once more together.

The mining company for which we were acting, is called the "Algonquin," and is composed principally of citizens of Detroit. Our locations were made, four in number, upon the waters of Flint Steel River, and we were now on the way thither to make preparations for the men who were to stay through the winter. Towards evening we entered the mouth of Flint Steel River, which is six miles east of the Ontonagon. Dragging the boat over the bar, and rowing it two miles up the stream, we landed. From thence to the locations is about twelve miles over a beautiful rolling country of sugar maple. The copper found here is chiefly native, and is enclosed in the trap rock. We brought away a piece weighing seven pounds, that lay in a vein near the surface.

On the 13th we were again at the boat, working out of the river. For several days there had been snow and indications of the close of the season. The snow was still falling as we proceeded down the lake after dark, with a view of reaching Elm River; but the water was calm, and the oarsmen were making good speed. A little after nine o'clock we passed the mouth of Misery River, a bleak and desert place without firewood, and some of the party fancying they saw a light at the old camp at Elm River, the boat was kept on her course. It was difficult to see the shore at the distance of twenty rods on account of the falling snow.

About half-past nine, a light puff came on from the north-west, which aroused the attention of Martin at once. "If the next one (says he) is stiffer than that, we must put about for Misery River." A sharp flaw followed his words, and the boat was put about. But it was scarcely before the breeze, when it came in short, irregular blasts, and the water became agitated. Martin was our oracle on the water. He said we must make the shore instantly, and the craft, bounding and splashing, was headed for a light streak that appeared to be a sand beach, but above which frowned a dark line like a bluff. Before she struck, the sharp, irregular waves combed freely over the sides and the stern of the boat.

"Charley, Patrick, Mike, and all hands, throw your oars and jump ashore!" Every man was in the water in a moment, holding her by the head. "Keep her stern off; heave, ho! heave, ho! Now she sticks. Throw out the luggage before she fills. Keep her stern off; heave, ho! Now she rests; take a line to that root." It would seem that not more than five minutes had passed since we were quietly moving over that water, from which we were now thankful to seek relief on land. The storm had already become a tempest, roaring through the woods and over the waves like a tornado. There stood the giant frame of Charley at the stern of the boat, the waves dashing over him, lifting and pushing her towards the shore; the others grasping her by the sides, assisted to work her further on, but she was too much loaded with water to be moved by main strength; Martin soon rigged the halyards into a purchase with two blocks, by which advantage she was drawn beyond the reach of the sea, that seemed to grow more angry as we rescued the boat from that element.

There is generally within hailing distance a birch tree to be found, and the ragged outside bark, that rolls up like paper in tatters, will burn at the touch of fire. No matter whether the tree is green or dry, or the day has been wet or dry, there is some side of a birch tree from which there can be pulled a handfull of these paper-like shreds, to kindle a fire. These, with a few small dead cedar limbs, will always with due care give the foundation of a camp-fire. But to be more certain, voyageurs usually carry a roll of peeled birch bark, the remains of some bark canoe, and this, broken and split intc strips, burns at once. Groping about among the balsams and pines that stood thick on the beach, no birch could be found. The roll in the boat had been washed out, and though found at last, was coarse and wet. The wind and snow which penetrated every nook and corner added to the difficulty of starting a blaze, and some of the party began to yield to the influence of cold and exhaustion, when we found a piece of dry pine board, and cutting it into shavings, had the satisfaction to see it flame up brightly at the root of a tree. A dish of hot tea revived every one, and at one o'clock the whole party were as sound asleep as ever, in a little hollow, back from the shore. But the storm raged on until the morning after the succeeding day, when we ventured to put ourselves before it, and reached Copper Harbor, sixty miles distant, in eleven hours without landing. As we passed the Eagle River, a number of people were seen along the coast where the spray still dashed over the rocks, in search, as we afterwards learned, of the body of Dr. Houghton, who with two of his men were lost there as the gale arose. It is remarkable that no more persons were shipwrecked on that dreadful night. A birch canoe with an Indian and his boy, and a white man, put out from Agate Harbor, and sailed in the height of the storm to Eagle Harbor, several miles. Other boats were exposed at various points, but, by seeking the shore in season, escaped the danger. Dr. H. had the misfortune to be opposite a forbidding coast, with rocks extending into the water, and shallow for some distance out. It was not his misfortune alone, but that of science and the nation. The boat did not, as it appears from the survivors, *capsize*, so capable is a well-built sail-boat of resisting severe weather, but was sent end over end, probably by hitting the bottom, while in a trough of the sea.

In September, a boat of about the same size made the passage from Isle Royal to Copper Harbor, direct across the open lake, with a bark canoe in tow, before a severe gale. A party of seven men, among whom was Mr. Hall of the New York survey, were on the island, and short of provisions. The vessel which was expected to take them off had missed the rendezvous, and they were driven to attempt the passage in their open boats. When fairly out on the lake, the wind which was fair increased to a gale, in which they gave themselves up for lost. About midway from the two shores the canoe and two men went adrift, and it became necessary to put about and take them again in tow. When it is considered how much the lug of a canoe impedes and endangers a small sail boat in bad weather, it will be regarded as a miracle of preservation that these men completed their voyage in safety.

I intended to give a brief notice of the mines now in operation, but have already made a much longer article, as I fear, than will suit a magazine reader.

The most extensive works are those belonging to the "Lake Superior Company," at Eagle River, under the superintendence of Colonel *C. H. Gratiot*. There were here about 120 workmen, and in September near 800 tons of ore ready for the stamping or crushing machine. This machine is a very nice piece of mechanism that works by water, and crushes ten tons of the rock in a day. The principal shaft, then seventy feet deep, was in a vein or dyke about eleven feet wide, one-half of which bears native silver in such quantities as to be an object without regarding the copper. Whether it is a true vein, or an irregular mass, I find geologists do not agree; but for practical purposes it is regular and extensive.

About four miles south-west from this, the "Pittsburg Company" are working a vein about four feet wide, which bears silver also, but its value is not as well tested as the Lake Superior Company's bed. Eagle River is only a brook coming down from the mountains, which a man may cross by two steps at low water. The shaft and pounding mill are about one and a half miles from the shore, and their landing is five or six miles east. At Eagle Harbor they have a saw-mill and many buildings. The celebrity of the mines, and the scarcity of places of shelter, have caused a great many persons to visit the spot during the past season. The superintendent and his assistants have, however, always shown visitors that attention and hospitality which could nowhere be esteemed more highly. About three miles east of Eagle River is the Henshaw location, not as yet much worked. On the west side of Eagle Harbor, at Sprague's location, I procured a handsome specimen of silver, which appeared to be abundant. On the east side is the Bailey location, not worked, but which is well spoken of. On Agate Harbor the "New York and Lake Superior Company" had sunk three shafts without hitting the metallic vein. The "Boston Company" have an establishment at the east end of the harbor. Within two miles, on the east, there are two veins, from one of which a piece of native copper, weighing about 400 pounds, was taken by Mr. Hempstead, and in the other a valuable sulphuret of copper has since been discovered. A vein of sulphuret is also known on the waters of Mineral Creek, a few miles west of the Ontonagon.

The "Massachusetts Company" have commenced works about a mile west of the extremity of Copper Harbor, where several veins, apparently rich, and said to carry silver, have been opened on the coast. At the Harbor the "Pittsburgh Company" have two shafts, from which they have taken several tons of the rich black oxyde. A mile east is a location of the "Isle Royal Company," under the charge of Mr. *Cyrus Mendenhall*, employing ten or fifteen hands.

There are probably now in the country 600 persons engaged in mining, as laborers, agents, clerks, superintendents, and mining engineers.

Communication is kept up with them during the winter, by a semi-monthly mail from Green Bay, taken on the back of a man by way of the Menominee River and the Anse to the post-office at Fort Wilkins. This does not allow the carriage of newspapers or heavy packages, but only letters. Although the winter is severe, it is so uniform that those who have tried it do not complain, and even pursue their journeys with more facility by land than they can in summer. If a road were open to Green Bay, the journey would be made in four or five days over a road which, once trod, would be perfect for several months. From the best information derived from mail carriers and gentlemen who have made the trip on snow-shoes, it is not an expensive route for a road.

I have spoken frequently of the fluctuations of the needle, and of its variations. The surveys in this region can be made only with the solar compass, or some instrument of that nature. The one used by Judge Burt, who has run all the township lines west of the Sault, is of his own invention. It is now made in England for exportation to this country. This compass is placed in the meridian by an apparatus always directed on the sun, and as it carries a needle, shows the variation every time it is set.

At the Sault, the regular variation was given 2° east, which, at every section corner on the town lines, is written with red chalk on the stake. At south-west corner section 19, range 35 west, T. 55 north, variation 7° 15′ east; 6 miles directly south, 5° 15′ east.

One mile north of south-east corner of T. 52 north, range 36 west, variation 5° 5′; one mile west, 6° 5′. At south corner of T. 52, range 37, variation 5° 15¼′ east; one mile north, 1° 10′; two miles west, 1° 35′; three miles further west, 8° 15′. At middle of south line of T. 51 north, range 40 west, variation 5° 35′ east.

For game we saw pheasants, or as some call them partridges, in great numbers, and also red squirrels. No turkeys, deer, or black squirrels. There are bears, moose, and reindeer; yet they are not numerous. There is also an animal of the wild-cat species, called a lynx, whose tracks we saw. For reptiles we saw none but a few feeble garter snakes. There are owls, mice, and rabbits in abundance. We saw no insects of consequence, except spiders, and these were sufficiently numerous to be troublesome. During the latter part of June and the whole of July, in the woods and low places, there are countless myriads of mosquitoes and sand-flies. They are said not to be troublesome on the coast.

Much of the comfort of a trip in this region depends on the outfit. Arrangements should be made for a supply of at least two pounds of solid food per day for each man, and a surplus for friends who are less provident.

The cheapest, least weighty and bulky, as well as the best for health and relish, are hard bread, beans, and salt pork of the very best quality. Tea, coffee, and sugar are in such cases not necessaries, but are, for the expense and trouble, the greatest and cheapest luxuries that can be had under any circumstances. To every two men there must be a small camp-kettle, and if in a boat a large kettle and frying-pan. In the woods, a hatchet to every two men, and a strong tin cup for each, with a surplus of one-half these articles to make up for losses. Knives, forks, and spoons disappear so fast, that two sets to each man will be none too many. Salt and pepper are indispensable for the game you may kill; and if there are a plenty of horse-pistols, a great many pheasants may be shot without much loss of time. But these are not to be taken into account for supplies.

A pocket-compass is necessary to each party. For a pack there is nothing better than a knapsack and straps, without the boards. Ordinary clothing is of no use, for it will disappear in a short time. The surveyors wear trousers made of heavy cotton ticking, and a sort of pea-jacket made of the same. This or medium cotton duck will stand wear, and although moisture comes through, the rains do not. It thickens when wet, and turns long storms better than any thing except oil-cloth. A supply of thick flannel shirts should be procured without fail, and flannel underclothes. A vest is unnecessary, and instead of suspenders the pantaloons are kept up by a broad belt, on which the tin-cup may be strung. A low, round-crowned, white beaver hat is much worn, but perhaps a light cap of oiled silk, made soft and impervious to rain, is better. For the feet, moccasins or light brogans made of good leather, and plenty of woollen stockings. In the wet season, cowhide boots, made of good but not heavy leather, and very large, but in the shape of the foot. A flint and steel for emergencies, and matches for ordinary use to strike a fire. Without something water-proof around them, the matches will acquire moisture in long spells of wet weather. If you carry a map case, they may be put in a second case, around which the map is rolled. A belt with a leather pouch and a buckle, to carry the hatchet in, is a very great convenience; for nothing is so likely to be lost as a hatchet. We were three days without one in very bad weather, having dropped it on the route.

Tents are not indispensable, but comfortable, especially along the shore, and in very warm weather when musquitoes are plenty.

A good, large, heavy Mackinaw blanket is beyond comparison the most necessary article to the voyageur and woodsman. With all these preparations, the lover of exercise and adventure may count upon as much enjoyment on a trip through the Lake Superior country as he will find at home. If he is badly provided, he will be inefficient and uneasy — will suffer many privations, and perhaps injure his health.

ON THE AGRICULTURAL INTEREST AND CONDITION OF OHIO,

Delivered in the Hall of Representatives at Columbus, Jan. 29, 1845.

[Ohio Cultivator, February 1845.]

CORRESPONDENCE.

To Charles Whittlesey, Esq.

Dear Sir:—The undersigned Committee were appointed at a meeting in the Hall of Representatives, on the evening of the 27th inst., to solicit of you a copy of your able and excellent address, on the subject of an Agricultural Survey of Ohio, and the importance of Agricultural improvements generally.

Your favorable answer will be thankfully received.

SEABURY FORD,
S. MEDARY,
J. RIDGWAY, } *Committee.*

January 31, 1845.

To Messrs. Ford, Medary and Ridgway, Committee, &c.

Gentlemen:—The manuscript copy of the address to which your note of yesterday refers is placed at your disposal.

If the striking facts which it contains, respecting the paramount importance of the agricultural interests of Ohio, shall serve to awaken and concentrate public opinion, I shall feel amply compensated for the labor of collecting and presenting those facts.

Very respectfully and truly yours,
CHARLES WHITTLESEY.

[Owing to the crowded state of our columns, we are compelled to omit the introductory portion of the address, in which the author gives a particular account of the soil and the farming of Hamilton county, where he was engaged in making an agricultural survey, the past year, under the auspices of the County Agricultural Society.—Ed.]

THE ADDRESS.

(The first paragraph is the conclusion of remarks on the county of Hamilton.)

I have stated that fifty years only have elapsed since the Miami region came under the axe and plough, and for much of that territory it may be said that it has not been cultivated over thirty years. I have given a particular statement of the kind of soil and subsoil which causes its fertility, and therefore it will be seen that nature has done as much for the Miami country as can be expected of her anywhere.

By personal examination, I find many tracts, and indeed entire farms, in this highly favored situation, that are so reduced as no longer to afford a reasonable profit, or even a living compensation for the labor and expense of cropping—to say nothing of the original cost or present value of the land; I mean that a man would not secure a good living, by working it, and paying the taxes, without a change in the system of cultivation.

This case is not a common one, but it is a common thing to see a farm that does not produce more than two-thirds of a crop. By this I mean, that the primitive capacity of the soil, uninjured by cultvation, as all soil should be, would, with the same labor, seed, and taxation, give a yield one-third greater. It is not necessary to confine ourselves in our expectations to this standard. Yet, in this country, most of our lands are, in their original state, good enough. We are not yet compelled, as the people of Flanders and England are, to *create* soil. We have it already furnished, of a good quality, and are not driven, as those people are, to devise methods of making *barren* land productive. It will be time enough to consider that question when we are pushed by population from our present happy position, where deserts are not known, to the sandy wilds, at the sources of the Arkansas.

The barrenness with which we have to contend is one of our own creating.

COUNTY AGRICULTURAL SOCIETY AND SURVEY.

It was with a realizing sense of these facts before them, and with the striking example of a most productive soil, occasionally broken down and made worthless within the life and remembrance of those who had enjoyed its original profuseness, that the farmers of Hamilton began to seek the ways and means of restoration.

They were aware that the first step to be taken was of an intellectual or mental character.

That it was necessary to set the minds of those who make farming a business, at work, not only by calling their attention to the fact of deterioration, but the causes and manner by which it has been brought about. If so many practical men can be brought to reflect upon the subject, an important point has been gained; for in our intelligent community the action of a multitude of minds directed to one object must result in something valuable. Next in consequence to the consideration of the subject is the mutual communication of the *results* or *conclusions* of these minds. After this information is collected and circulated, it becomes the property of all; and if it does not bless and improve them, the fault is clearly their own. But while it is not in their possession, they may perhaps be called to an account for their ignorance, but certainly not for the wilful abuse of knowledge.

The initial steps to encourage investigations of

this kind were taken by the legislature in providing a general law for county Agricultural Associations in 1839.

Although this law is in many respects thought to be capable of improvement, it proves to be sufficient for the organization of societies, and under it the Hamilton County Society was formed. In this manner, something visible and tangible was constituted to attract the attention and respect of the public, and its published proceedings will show what has been accomplished.*

The survey or examination of the farms of the county may be regarded as one of its most important movements. This having been completed in an imperfect manner, the society, in their corporate capacity, have recommended, by a formal petition to the legislature, the extension of similar surveys to all the counties of the State.

A committee or delegation was appointed for the purpose of presenting this petition, and suggesting arguments in favor of the scheme.

It is composed of Messrs. Brown and Flinn, members of the lower house, from that county, of Mr. A. Randall, a director, and myself.

It is as a member of that commission, and by its authority, that I present the subject this evening in this form. There is, I am aware, among farmers, an aversion to what is called Book Farming, and to book knowledge on the subject of farming.

This is not strange, because soils, climates, and circumstances are so diverse, that what is true of one place may be wholly false and erroneous in another. The fault in such cases is not, however, in the facts, but the application of them. The farmer who takes

* A part of this Report relating to wheat, as published in the *Ohio Statesman*, February 3d, 1845, is attached to this Address.

up a book written upon cultivation in Flanders may obey its direction ever so implicitly, and may not only lose his crop, but injure his land. It requires discretion in the application of knowledge in farming as well as in medicine or any other calling. The soil is a great chemical laboratory, where organic changes are continually going on. The physician, also, deals in compounds of a chemical character; most of them, when misapplied, are injurious and even fatal. It is not perhaps necessary that the doctor should be a chemist, and be able to *combine* and *originate* all the medicines he uses; but it is necessary that he have a certain degree of information respecting their nature, origin, and effects, or he is an unsafe man to have care of our health and life. He must at least know the ingredients and their properties. Until he has this knowledge, he is incapable of exercising an intelligent discretion in the application of remedies, and it will be an equal chance whether he kills or cures. A certain portion of this same knowledge is *advantageous*, though not perhaps as *necessary*, in farming. Soil is supposed to be formed entirely by chemical action.

ORIGIN AND COMPOSITION OF SOILS.

The earth, as it came naked from the creation, was destitute of vegetation. It is supposed to have been a mere mineral mass, containing, it is true, the powers of germination, and a feeble ability to support plants when germinated. When the first plant was grown, it fell into decay, containing within itself various elements, and combinations of elements, such as oxygen, carbon, hydrogen, nitrogen, and various acids, such as phosphoric, acetic, sulphuric, and earths and alkalies, such as silex, magnesia, potash, soda, and lime. All these have been extracted from the air, the earth, and the waters of heaven, and, by the mysterious power of vegetable life, fashioned into a beautiful object.

The stalk, leaves, and fruit of the plant, with so many substances and compounds of substances, are dissolved, and perish. But the matter of which it is composed does not perish, it only seeks new combinations. In the earth on which it rots, there are alkalies, earths, and oxides, by which the acids and gases of the decaying weed have, by the law of nature, a strong affinity; a chemical desire so powerful that they immediately reunite. When burnt lime is exposed to the atmosphere, we observe that it soon acquires carbonic acid, and this gas, combining with the caustic lime, causes mortar to harden by age. This is an example of chemical affinity.

The operation is silent, and apparently weak and trifling. But it is by a knowledge of this property of lime that mortars are made, and by means of mortars that edifices, aqueducts, and fortifications are constructed as solid and lasting as the natural rock.

In my opinion, there is no material difference in the mineral constituents of soil in its primitive state and the subsoil or earth beneath it; that the reason why the surface matter is more fertile than that at the depth of 10, 12, or 18 inches, is the *chemical change* that has been wrought by vegetation, air, heat, frost, and moisture. The solid particles are furnished by the earth; the acid and gaseous materials by the plant; and these being brought in contact, a lively chemical action commences. By this means, the soil which is naturally red, yellow, or white, gradually becomes blacker; where it was compact, if the operation is well effected, it acquires porosity and looseness.

This being the *manner* in which soil or vegetable mould was originally separated from subsoil, or mere earth, we have only to imitate nature to produce it ourselves.

Such is the theory of manures. There are soils that are radically deficient in the alkaline bases; for these, vegetable manures would be of little avail, because some of the chemical elements are wanting. There are others where the alkalies and salts are abundant, but they want vegetable matter. Furnish it, and a luxuriant crop rewards the husbandman. It is, therefore, upon a *judicious mixture* of these substances that fertility depends. The exposition of these phenomena is *book knowledge*. The application of chemistry, showing how soils are constituted, and why certain ingrèdients are necessary, is *book farming*, or science brought to the aid of labor.

The experience of one is made accessible to every one else, by the means of *printed books*, and the misfortune is that they are not more numerous and more cheap.

A farmer, with a soil already sufficiently calcareous, has no need of lime, and if he expends his money or work in carting it to his premises, will gain nothing by the operation, and may produce a permanent injury to his soil. Applying lime to every kind of land would be like giving calomel in every disease. It might with as much propriety be said that the book learning, which shows the powers and benefits of that medicine, was useless or unimportant, as to speak thus of those treatises on agricultural chemistry which explain the nature of soils.

PROCESS OF EXHAUSTION.

The first settlers of the West appear to have regarded our rich lands as possessed of inexhaustible fertility. The people of the new counties of Ohio, where the process of cultivation has not been of long duration, still take little interest in agriculture as a science. Those who perform the severe labor of clearing the land are impatient to receive their reward, and apply their remaining energies to the work of drawing from the soil the most rapid succession of crops. The decline is so gradual as to be imperceptible for a short number of years, and so long as the bounty of nature holds out, her resources are drawn upon freely.

But the old settlers, who have survived half a century of active life, are enabled to *compare the extremes*, and to them the contrast between the primitive richness of their farms and their present power of production is capable of being observed. Those who remember when corn-land produced seventy-five bushels per acre, and still live to see the same land, with the same labor, give only forty bushels, *realize* the difference between a state of exhaustion and a state of original vigor. This difference being taken from the *profits*, and not from the entire *product*, becomes still more striking.

The *fact* of depreciation is, therefore, well established as a matter of evidence, but the *manner how* it is brought about is not always so well understood. By throwing light upon this sinking process, the abstract idea will become more sensible, and assume a prominence in the mind, equal to its importance in practice.

Only about fifteen per cent. of the matter of the western soils produces any direct effect upon vegetation. About eighty-five per cent. is mere sand and clay, and only serves to retain moisture, and supply a foundation or basis for the plant. Of the fifteen per cent., there is in the best soils an average of ten or twelve per cent. vegetable matter, but only about one-half of this is in an active state, say six per cent.

There is from one to three per cent. of lime in the state of carbonate, sulphate or phosphate. The soil actually contains a minute portion of potash, for we find it in the ashes of timber, and it must come from the earth. So plants and trees contain magnesia and soda, and sometimes the oxides of iron and manganese.

The iron is generally appreciable in quantity, and makes its appearance in the analysis, but it is difficult to detect the potash, soda, and magnesia, the quantity is so small. All the valuable mineral constituents amount to only three per cent., and the vegetable to *six*, making *nine* per cent., from which all the earthy supplies of vegetation are to be drawn.

It is not necessary to extract all the materials of this nine per cent. of the soil, in order to render it unfruitful, or even to exhaust one of them—for if we *diminish them, or one of them, so as materially to change their relations*, we have effected a disorganization of the soil.

The depth stirred by the plough is ordinarily *four* inches, sometimes five and even six inches. From this six inches of depth, or from nine per cent. of it, we draw annually of hay, grain or corn, from two to four tons of vegetable substance, or say on an average 6000 pounds. How great a portion of this product is derived from the atmosphere, and how much from the earth, is not a well-settled point. But if one-half is taken from the soil, it amounts in thirty years to 90,000 pounds, or forty-five tons of its very life-blood and sustenance.

The weight of a covering of earth, measuring six inches in depth, will vary from 1000 to 1280 tons per acre, and nine per cent. of the same to 90 and 115 tons, of which the forty-five tons taken up by the plants, in thirty years, amounts to fifty and thirty-three per cent.

I give this more by way of illustration than as well determined proportions, although my opinion is that they are not far from the truth. It is therefore easy to perceive *how* a soil is exhausted by cropping, and to realize that what ruins it in thirty years must do one-thirtieth part of the injury in one year.

PROCESS OF RESTORATION.

It also shows that, as the decay is comparatively slow, requiring time and continual cultivation to effect it, so the process of restoration cannot be brought to perfection at once, but will likewise require the lapse of time. This follows from the chemical action which is necessary in order to produce a change in the vegetables and alkalies present.

If the vegetable part is most deficient, and we resort to the usual mode of spreading manure upon the soil, it requires some months for this action to commence, for the decomposition to be effected, which precedes the new compositions that are to be formed. It may require years for the formation of all the compounds that successively appear in the soil after the application of good manure. This merely verifies a general rule of nature, that the reverse process of restoration is not more rapid than the direct one of depreciation. And this principle, well considered, impresses the fact indelibly upon the mind, that it is easier to *maintain* than to restore. That it is not only easier, but more profitable, to preserve a soil in its original strength, enabling it to produce its maximum all the while, than to suffer it to run down, lose the product, and then restore it, is a position that does not require an argument. There are methods of manuring which are more rapid than others, and more profitable, but the maxim I have just laid down should be remembered, that quick and powerful stimulants are soon themselves exhausted, and cease to operate.

It would be too tedious, if our discourse was prolonged so as to present in detail the theory of vegetation and conversion of manures, the analysis of various soils, and of the various vegetables that are produced upon them. The examination of this subject would show a correspondence between the composition of the soil, and the nature and luxuriance of its productions. It would explain why some plants, as clover for instance, when turned under as a green crop, produces more fertility than buckwheat or corn-stalks.

These investigations are purely chemical, and have been particularly brought about by the influence of agricultural associations.

ASSOCIATIONS—"THE LONDON BOARD OF AGRICULTURE."

The most noted society of this kind, and one which may be regarded as laying the foundation of British husbandry, now reduced to a science, was the "London Board of Agriculture," established by an Act of Parliament, May 17, 1793, and furnished with £3000 a year from the Treasury. At the close of the American Revolution, the island of Great Britain was estimated to contain 60,000,000 of acres, exclusive of cities, roads, lakes, &c., of which only 30,000,000, or one-half were in cultivation.

Sir John Sinclair had at his own expense travelled in Flanders, Germany, France, and generally through Europe, and observed, that for their surface those countries were producing much more largely than England.

He proposed an inquiry into the causes of the striking difference that existed in the agricultural condition of the Island and the Continent, and broached the project to the ministry. He was told, that whatever related to commerce and acquisition of territory, to the army and navy, would meet with a ready support in the cabinet, but they never had, and probably never would, bestow that attention upon agriculture.

It was expected that the ministry regarding husbandry as a mere handicraft, upon a level with the excavation of a cellar, or the raising of an embankment, would not descend to consider the proposed scheme. Although George III. had bestowed some attention upon practical farming, the ministry supposed it would take care of itself, requiring only the requisite number of laborers. They regarded the muscles of the human arm, the strength of horses, the plough, the mattock, and the spade, as constituting the sum total of agriculture, and that intelligence and science had no more to do with its improvement, than it had with excavating the cellar, or piling earth into the embankment.

Mr. Pitt was not really for the project, unless the House manifested some desire for its adoption. Mr. Dundas took favorable ground, and after much persuasion, and even importunity, Sir John Sinclair at last attained his object. In arguing the question before the officers of government, he proposed to gain *six* principal ends, of which I think *three* are applicable to this country at this time.

1st. The Central Board would be a general *magazine* of agricultural knowledge. 2d. It should be their duty to collect and circulate this knowledge. 3d. As a part of this duty, to cause a *survey* of *England* to be made by counties, giving a statistical view of its present state, wants, and ameliorations. The Board being organized, composed of the principal councillors of State, some eminent clergymen, and thirty members, their first business was the collection, by local agents, of the statistics and agricultural condition of the kingdom, which was effected by counties, and printed in about two years.

They next laid the foundation of Agricultural Chemistry, by procuring from the father of chemical science, Sir Humphrey Davy, a course of Lectures and analysis.

There had been associations for the benefit of agriculture in the kingdom before. In Scotland, as early as 1723, the society of "Improvers in Agriculture" was organized, and embraced, for a time, many valuable members. In 1749, in Ireland, the Dublin Agricultural Society was formed and received a grant of £10,000 from the Irish treasury, for the promotion of its objects. The "Bath and West of England" Society arose in 1777, and the Highland Society of Scotland in 1784. But all these associations were limited, comparatively powerless and temporary. The London Board was composed of men whose interest in the cause was intense, and by a connection with the government, they were enabled to command means to accomplish their designs. The consequence was that, in 1796, the Board report that, 22,350,000 acres of the land had been reclaimed, from a waste or unproductive state, and added to the wealth of the nation. Its value was estimated at £905,215,500 sterling.

The Board demonstrated that the fears then prevalent of over population were without foundation, as they might and had been overthrown by over production of the soil.

This striking result in England was not all the advantage resulting to that country and to other nations, from the labors of the Board of Agriculture, which continued until 1819. That Board gave rise to works especially devoted to analysis of soils, grain, straw, and all vegetable substances; attracted the attention of the agricultural world to the subject of improving soils; engaged chemists in the work of examination of manures, and laid the foundation of vast improvements.

Agriculture is an art which has never been known to recede, but always to advance and improve.

The Egyptians, when they cultivated the valley of the Nile, had their ploughs, their yokes for cattle, and their thongs of leather, to attach the team to the plough. But their contrivances were rude and cumbrous. The Greeks of the time of Homer were practical farmers, and had improved upon the implements of the Egyptians.

So the Romans of the days of Hesiod had advanced upon the Greeks, the Germans of the Rhine upon the Romans, and the English of the last century gave agriculture an impulse over the Flemish and German standard. In America, this progress is in my opinion to be extended, and another step taken towards perfection — particularly in implements.

IMPORTANCE OF AGRICULTURE — STATISTICS.

May it not be said that, hitherto, and even now, the general government and the States indulge in too much indifference respecting the advance of agriculture. It is, in the language of the petitioners whom I represent, the basis of every other interest. It is the principal and reliable source of taxation among the States. In Ohio the revenue is derived almost entirely from real estate. The commercial interest represents merely the surplus of the agricultural. The property engaged in manufactures is limited, compared with that invested in the soil.

The census of 1840 gives $32,201,263 as the capital engaged in forwarding, in the trade of merchants, lumber and butchering.

The capital engaged in manufactures is represented as $14,905,257.

The products of the soil in Ohio, for the same year, were as follows:

Wheat,	16,571,661	bush.	at 60 cts.	$9,942,996
Barley,	212,440	"	" 40 "	84,976
Oats,	14,393,103	"	" 20 "	2,878,620
Rye,	814,205	"	" 40 "	325,682
Buckwheat,	633,139	"	" 30 "	189,941
Corn,	33,668,144	"	" 30 "	10,100,443
Potatoes,	5,805,021	"	" 20 "	1,161,004
Tobacco,	5,942,275	pounds	3 "	178,268
Hay,	1,022,037	tons	$6	6,132,222
Hemp Flax,	9,080	"	$100	908,000
Hops,	62,195	pounds	10 cts.	6,219
				$31,908,371

The above includes only the crops proper for 1840, and not for 1844. I take the produce of the former year, because we have the official report as a basis as to the quantity, and I have given the prices below, rather than above, the market. The wheat crop of 1844 was probably less than that of 1840, on account of a bad season, just as it arrived at maturity. But the general increase of production in Ohio, over that above given, may be safely put at *one-fifth,* or 20 per cent.

There are some important additions to be made to this table of articles, the result of agriculture, not properly termed crops.

Products of	Orchards,	for 1840,	$475,271
"	Dairies,	"	1,848,869
"	Gardens,	"	97,606
"	Nurseries,	"	19,707

Wine,	11,524	gallons at $1		$11,524
Silk,	4,317	pounds " 5		21,585
Wood,	272,529	cords " 2		545,058
Sugar,	6,363,386	pounds " 4	cents,	254,535
Wool,	3,685,315	" " 30	"	1,005,594
Wax,	38,139	" " 25	"	9,534
				$4,289,283

This sum, in addition to the value of crops, gives $36,197,654.

To this should be added the annual increase in value of animals, to wit:

430,527	Horses and Mules,	at $50	$21,526,350
1,217,874	Cattle,	" 20	24,357,480
2,099,945	Hogs,	" 3	6,299,835
2,028,401	Sheep,	" 1	2,028,401
			$54,212,066

If the annual proceeds of the live stock of the farm is put at *one-fourth* the value, and my estimate of that value is correct, the yearly product would have been, in 1840, $13,553,016.

This, item united with the annual value of crops and agricultural products, makes $49,750,670.

But there are many things not included in this calculation, such as the value of pasturage, straw, turnips, poultry, feathers, &c., which would swell the sum considerably.

And although the crop of wheat was this year greatly injured, and the crop of corn rather light, and the product of orchards diminished, there must be for 1844 a material increase in production over 1840.

I think it would be safe to add 20 per cent. on that account, to the estimate just given, which would give for the agricultural products of Ohio at this time $56,990,197.

This sum, it will be remembered, does not fully represent the agricultural interest or capital, but the gross production of that capital.

There are upon the tax-list of this State, for 1842, 20,260,526 acres of land, of which the value per acre cannot be less than five dollars.

It is not easy to make an estimate of the number of farm houses, buildings, and implements in the State, or their value.

In four rural townships of Hamilton county, which in 1840 contained a population of 7411, there are but 243 houses returned for taxation, or about one to thirty inhabitants.

By law, buildings below a certain value are not put on the duplicate, which deprives us of information from that source.

By taking the average price of land in Ohio, as established by the Board of Equalization, $3,68, we can find the value of lands as they stand upon the grand levy, and deducting this sum from the value of houses and lands, we have $8,835,492, representing buildings, principally farm houses. But the method of assessment, like that of lands, places these buildings far below their real value. And it should also be remembered, that barns and out-houses are omitted, and manufactures included.

I think it reasonable, however, to multiply the amount stated on the duplicate by *four*, and call the product the real value of farm houses, barns, out-houses and implements, which will be equal to $35,341,968, or a little over $100 to each individual engaged in agriculture.

The lands of Ohio, aside from town lots, at $5,00 per acre, are worth $101,302,630. Of the live stock,

I regard *one-fourth* as annual increase, and three-fourths as capital producing this increase. The agricultural investment in Ohio may be considered as the aggregate of lands, buildings, implements, and three-fourths of the stock.

My estimates are, of course, only rough approximations to the truth, but these three items make a gross sum of

¾ Stock, - - - - - -	$40,647,855
Lands, - - - - - - -	101,302,630
Houses, &c., - - - - -	35,341,968
	$177,292,453

The annual product of which, according to the foregoing estimates, is $56,990,197.

The united capital of merchants, forwarders, butchers, of lumbermen and all manufacturers, we have given at $59,106,520, and adding one-fifth for increase to the present time, it gives $90,927,824 capital in trade and manufactures, against $177,292,-453 invested in agriculture. In other and more commercial States, the disproportion would not be so striking, but still, throughout the United States, the agricultural interest, measured by dollars and cents, or by the numbers engaged, or its importance to the nation, stands above *any* other, if not above *every other*, department of investment and industry.

In 1840, when the population of Ohio was 1,519,-467, her agricultural laborers numbered 272,599. At the same time, the persons engaged in all the other business, callings, trades or professions, amounted to only 84,458. If the farmers have increased in proportion with the population, they now number about 323,000.

GEN. WASHINGTON'S OPINION — GENERAL GOVERNMENT.

Has the legislation of the country, in favor of this overruling interest, been proportioned to its magnitude?

We have witnessed the extreme attention of the National Congress and of the State Legislatures to the subject of commerce. From 1789, to this day, the Federal Government has not only been interested, but agitated, and that almost without cessation, by different schemes for the promotion of manufactures.

In the last annual address of President Washington, wherein he lays down the great principles that should govern our statesmen, he does not forget to recommend the protection of agriculture, as well as trade and manufactures, in the following terms:

"It will not be doubted that, with reference to either individual or national welfare, agriculture is of primary importance. In proportion as nations advance in population, and other circumstances of maturity, this truth becomes more apparent, and renders the cultivation of the soil more and more an object of public patronage. Institutions for promoting it, grow up, supported by the public purse, and to what object can it be dedicated with greater propriety?

"Among the means which have been employed to this end, none had been attended with greater success than the establishment of Boards, composed of proper characters, charged with collecting and diffusing information, and enabled by premiums and small pecuniary aids to encourage and assist a spirit of discovery and improvement. This species of establishment contributes doubly to the increase of improvement by stimulating to enterprise and experiment, and by drawing to a common centre the results every where of individual skill and observation—and spreading them thence over the whole nation. Experience, accordingly, has shown, that they are very cheap instruments of immense national benefits."—*Journals 3d and 4th Congress, page* 69.

What has been done in pursuance of this recommendation, relating to the soil, and what has not been done respecting those that refer to the other great callings? Is there less constitutional power to favor the greatest interest of the country than for those which are subordinate? When has there been in Congress any direct legislation, avowed and intended to protect the farmer, that was not subordinate and subservient, or at least incidental, to the other branches? The capital invested in commerce in the United States,

is estimated at	$262,000,000
In manufactures at	267,000,000
For both . .	$529,000,000

How many laws have been passed to protect manufactures and encourage commerce? The annual product of the United States, in the article of vegetable food for man, without including any thing else, is $624,518,510. The value of horses, mules, cattle, sheep, and swine, is $640,000,000, and adding all real products of the land, we may safely put the grand aggregate at $1,000,000,000, or double the capital employed in trade and fabrication. Has this subject been so long overlooked because the farmer needs no encouragement?

The 323,000 men, engaged in cultivation in this State, produce, according to my estimate, $56,990,197.

Their wages, at $15 per month, would amount to $58,140,000.

The interest upon the property invested, $177,292,453, at six per cent., is $10,637,547. Where is the general profit on agricultural investments? It is evident, without entering into a calculation, that if farms and farming paid a fair interest upon money, the capitalist would invest money in the business. But he does not—he prefers trade or manufactures; and experience shows that he realizes more from such an employment of his money than he would by tilling the soil.

We have then this anomaly, the most important occupation of man, that in which more persons are employed, will only bear an investment of labor, not of money capital. We find the legislation of the country active, sensitive, untiring, to make this money capital yield a high profit, but find no statute of Congress, having for its object a direct, exclusive, particular application to the profit on this great labor capital.

Much has been done for it under other names, as an incidental affair; but have we not imitated too closely the indifference of Mr. Pitt and his colleagues, in 1793? What encouragement does it require? The general government having neglected the subject in a great degree, this question is more directly applied to the States.

But to answer it, consultation and reflection are both requisite.

DUTY OF THE STATE GOVERNMENT.

The diffusion of cheap practical books would give agriculture a character and an impulse. It would tend to make the calling more elevated, because more intellectual.

Agricultural schools, where labor and learning mingle their benefits, is another mode of advancing the object.

A more full analysis of all kinds of soil, and in connection with that work, the analysis of all vegetable products, to discover the mineral and gaseous constituents, they take up from, and give back to the earth.

Sir Humphrey Davy, Chaptal, and Liebig, have done much, and the American chemists extended the subject, but without a special devotion to this single branch of chemical investigation, we cannot expect a full knowledge of it.

The subject of manures, not only in regard to their *effects*, but their *cheapness*, requires more examination and experiment, and every country must make them for itself. Results obtained in Germany may not answer in Ohio.

Agricultural *statistics* are indispensable, and in this branch the general government did the country much service in the completion of the census of 1840.

A general *interchange of the experience of practical farmers* appears to me, for its expense, the most useful of all modes of advancing the state of our tillage. There are men, who with the same soil, labor and expense, produce double the surplus of others. The former class of persons, by a trifling but incessant application of some kind of manure, maintain the soil in its original strength. Their crops are sure and heavy. Another drains the soil, until the crop falls away visibly. He cannot cease to till it, for he would starve. He cannot manure it all at once for want of means. If he goes on plowing, sowing, and reaping, as usual, he falls behind annually in his expenses, the tax-gatherer crowds him, and the surplus of his farm is no longer to be found.

Let every farmer in the State relapse into this condition, and where is the wealth and happiness of our people?

Let every farmer strive to pass this fatal point, where the cost of cultivation is just equal to the proceeds, and what a vast amount of surplus would be at our command.

I venture to prolong this discourse to give an example: There are not far from 1,700,000 acres of land in cultivation, in wheat, in the State of Ohio. In Hamilton county, the average yield of upland wheat per acre is 16½ bushels, and the average price at market, for the past four years, is 67½ cents. The ordinary cost of cultivation, harvesting, &c., for wheat, I fix at from $7,25 to $7,50, or say the value of 11 bushels at 67½ cents, which is $7,42.

If the yield is no greater in other parts of the State than in Hamilton county, there is a surplus of only 4½ bushels, or $2,70, which is not equal to the interest upon the land, implements, and stock, required for its cultivation.

Our soil is capable of producing from twenty to twenty-five bushels of wheat, and with extraordinary attention thirty bushels. But with merely good farming, every where introduced, the same land and the same labor will add five bushels per acre to the crop of Ohio, — or 8,500,000 bushels, of the value of $5,737,500—this, without including other crops.

Now, I find by the current report of the Auditor of State, that the property of the State pays a tax of $2,340,663, for all public purposes, and this assessment is considered as unusually heavy, almost oppressive.

According to my estimate, the wheat growers of the State alone might, by careful cultivation, with the same quantity of land, enlarge their surplus to more than double this sum.

In 1843, the wheat crop of the United States was estimated, by Mr. Ellsworth, at 100,310,856 bushels, of which Ohio furnished about one-fifth, or 18,786,-705 bushels.

The average yield of wheat in England is twenty-eight bushels per acre, in the United States not to exceed fifteen, in Ohio seventeen. It would be profitable, in this State, to aim at twenty-five bushels, and easily practicable to attain to twenty-two, thus adding over one-fourth to our surplus in wheat.

This subject might be amplified and presented at much greater length, but I have already become tedious and close at this point.

WHEAT CROP OF HAMILTON COUNTY.

In passing through the county, it was my practice to note down the remarks of each individual respecting his farming operations; and this report will consist, in a considerable degree, of those remarks, arranged under their respective heads.

It was an object to obtain the yield—the time and manner of sowing—the system of manure—time and manner of harvesting of each crop; and also to ascertain the kind of timber which grew upon the soil—the length of time in cultivation—rotation and number of crops, and such other information as might be suggested at the moment.

We know of no better authority on these subjects than the farmers themselves, and do not feel ourselves competent to improve upon their opinions, statements, and suggestions, and therefore merely report, in as clear and methodical a manner as we can, the relations which were given: There is no doubt of a direct connection between the constitution of a soil and the timber it produces; and from this we may deduce a connection between timber and crops. The heaviest crop of wheat is found on land having sugar tree and oak as the principal timber, as will appear from the following classification:

Sugar and oak,	10 cases,	average	crop	18,40 bs.
Sugar and beech,	19 "	"	"	17,52 "

Beech and oak,	7	cases,	average	crop	17,14 bs.
Sugar, oak, hickory,	5	"	"	"	16,66 "
Oak,	21	"	"	"	16,00 "
Hickory,	10	"	"	"	14,50 "
Beech,	12	"	"	"	14,33 "
Miscellaneous,	30	"	"	"	15,50 "
General average of	127	"	"	"	16,50 "

The mixture of sugar and oak appears to give the best yield, sugar and beech next, and beech and oak the next. If an analysis of these soils and of the wood most congenial to them was carefully made, we should probably discover, not only a resemblance between them and their timber, but between the timber produced and the grain and straw of wheat.

The statement of an average yield of wheat enables us to ascertain the number of acres cultivated in wheat in 1839, when the official return gives 213,815 bushels, for the product of that year. If that year gave an average crop, the quantity of ground then in cultivation in wheat was 13,029 acres.

The crop since 1840 has been more uncertain in its yield, price lower and demand less, from which we conclude that there has not been an increase in the production of the county since that time.

MISFORTUNES TO WHICH THE CROP IS LIABLE.

The record we have given, of the remarks made upon wheat, shows that there are three principal evils to which it is subject, viz: rust, fly, and freezing out. The average loss of the crop by these and other causes is once in four years a total failure.

EVILS AND SEED.

The farmers differ in regard to which of these is the greatest enemy to wheat; six of those speaking upon the subject regard rust as the difficulty most to be avoided; nine consider freezing out the most injurious, and nine suppose the fly to be the worst enemy.

In regard to insects that prey upon crops, we have a separate notice of them, to which we refer, merely calling the attention of the reader, here, to the worms observed by Mr. W. W. Cary (30), and Mr. Thorndyke Keller (107), with a view to farther discoveries.

Respecting the quantity of seed, the farmers of Hamilton vary from three to five pecks. Mr. Smethhart (99) considers three pecks to be better calculated to insure a full crop than more. In Flanders, two Winchester bushels are sown to the English acre, which is nearly eight pecks of our measure; in England, two and a half to three and a half bushels.

The richness and depth of the soil has so much to do with the quantity of seed that it is difficult to establish a rule upon this subject. In England, however, where the dibbling process has been tried and the seed planted in regular squares of six inches on a side, and a seed in the middle of two sides, the quantity is reduced to about one and a half bushels to the acre, and the yield increased many fold.

TIME OF SOWING—PRICES.

Among the farmers who gave answers to our inquiries about the time of sowing, four put in their wheat in August, four in October, nine in September, and one in November; of those who prefer September, six sow during the first ten days, one by the first day, and two in the latter part of the month.

Those who cover the seed in the first week expect to harvest between the 25th and 30th of June.

The following table exhibits the price of wheat at Cincinnati, every six months, for four years past:

1840, July 1st,	56	cents per bushel.
1841, January 1st,	58	" "

1841, July 1st,	73	cents per bushel.
1842, January 1st,	1,06	" "
" July 1st,	55	" "
1843, January 1st,	50	" "
" July 1st,	85	" "
1844, January 1st,	75	" "
" July 1st,	60	" "

Average price per bushel, 67½ cents.
Average in January of each year, 72,22.
" " July " " 65,80.

The method of tilling is very various. In new ground, for the first crop, the seed is generally scattered over the fresh surface, and harrowed in without plowing. The rooty condition of the ground frequently prevents the use of the plow.

In old ground, among farmers, the practice most in vogue is to summer fallow and harrow in upon the sod. In fields that are clear of stumps and stones, and very few farms in Hamilton county are troubled with stone, the plow lays over the sward, very uniform and flat. It is thought that ground made too mellow and fine is more liable to winter-kill the wheat than such as is in the state of clods, provided they are dead and rotting.

Some cases will be observed in our abstract where clover has been turned in and wheat sowed upon furrow immediately, and also where turf land has been broken up a few weeks before seeding.

In land where plowing can be done so perfectly, as to turn over all the soil, it is not inclined to send up grass and weeds; this practice of sowing soon after breaking up is undoubtedly advantageous. It allows the growing plant a chance to absorb more of the volatile portions of the decomposing sward.

Upon the whole, this region cannot be said to be well adapted to the culture of wheat.

UNCERTAINTY OF CROPS.

A crop which is liable to be cut off once in four years is too uncertain to be raised with profit for sale, although every good farmer would continue, even under this disadvantage, to sow enough for his family use. It is much less sure, and also less abundant in yield, than formerly, when the country was new.

This seems to indicate a change of climate, as well as a change in the condition of the soil. The effect of cleared land, and of cultivation, is to cause a greater number of thawings and freezings, which are each of them injurious to the root, and when land becomes heavy by constant use, the effect of frost in that way is different and more destructive.

The country is filled with the fly, and there seems to be an increased disposition to rust. Very early sown wheat generally escapes the rust, but appears to be more exposed to the fly.

The Alabama wheat has hitherto escaped both, and produces a fine yield of good wheat. It has, however, not been tried long enough to pronounce upon its ultimate success. We should expect a grain from the South to mature earlier, and consequently do better, provided the winters did not injure it.

VARIETIES AND COMPOSITION.

A kind of wheat, called Virginia wheat, has been tried with that view, as we are told, and gave a good crop, but we have no personal knowledge concerning it.

Mr. Frost (64), of Crosby, has tried Saxony wheat, and considers both it and the "blue stem" as before our common kinds, and even before Alabama.

The fact noted by Mr. Brown (65), that the fly which infested the timothy part of his fallow, did not the clover, is worthy of remembrance.

In England there are forty-two varieties of cultivated wheat, winter, spring, and summer; but as yet they are not well classified in botanical order.

The Romans of the days of our Savior sowed the common red and the white wheat in general, but in moist situations they made use of the bearded.

Sir Humphrey Davy's analysis of wheat shows that the spring wheat contains the largest amount of gluten, and the Farmer's Encyclopœdia intimates that it is the most nutritious.

	Gluten.	Starch.	Insoluble matter.	Total.
English wheat, .	19	77	4	100
Sicilian " . .	20	75	5	100
Spring " .	24	70	6	100
Blighted " . .	13	52	35	100

The substances of a mineral kind, drawn from the earth, are, according to the analysis of *Sprengel*, for 1000 ℔s. of

Wheat and straw, per acre.	Wheat.	Wheat straw and Wheat.	
℔s.	℔s.	℔s.	
285	225	20	Potash.
337	240	29	Soda.
816	96	240	Lime.
186	90	32	Magnesia.
296	26	90	Alumina and iron.
9010	400	2870	Silica.
161	50	37	Sulphuric acid.
550	40	170	Phosphoric acid.
100	10	30	Chlorine.
117,41	11,77	35,18	Weight of ashes.

If we call the weight of straw per acre 3000 ℔s., which is near the average, and allow sixteen bushels of grain, or say 1000 ℔s. weight, we shall have 4000 ℔s. as the vegetable product, which leaves 117,41 ℔s. earthy residue.

This result is not strictly exact, and the grain of Germany, where Sprengel operated, may not be composed exactly as our own. The most striking fact contained in this table is the discrepancy between the amount of lime in the kernel compared with the stalk.

TIME OF HARVESTING.

Having given the principal details of our observations among the cultivators of the soil at home, we add some considerations from the experience of farmers in other countries, particularly in relation to the state of wheat when it is cut.

In 1840, Mr. John Hannum, of North Deighton, in Yorkshire, England,* made several experiments upon the relative value of wheat, cut at various stages of ripeness. It should be borne in mind that the harvests of England occur from four to six weeks later than in Southern Ohio:

Specimen	No. 1,	cut green,	August	4,
"	No. 2,	cut raw,	August	18,
"	No. 4,	cut ripe,	September	1,

The green specimen had not begun to turn yellow; the chaff adhered to the kernel, which was green, soft, and full of milk, although perfectly formed. No. 2, or the raw, was quite yellow from the roots about one foot upwards, and the whole stalk, though apparently green, was seen to be, upon close examination, of a yellowish tint. The ears were open, the chaff yellow

* Farmer's Encyclopœdia, article wheat, p. 1124.

and green, and the grain still soft and pulpy, with some fluid matter in the kernel. The specimen No. 3 was ripe, without being dead ripe or brittle, but in what is called harvesting order.

	bushels.		lbs.	lbs. straw.
The ripe gave	30	of	60	2,688
The raw gave	30,1307	"	"	2,352
The green gave	26,1356	"	"	2,737

The straw of the green parcel was, as we see, about 385 lbs. heavier than the raw, and 49 lbs. heavier than the ripe. The price per quarter of 8 bushels, showing the quality in market, was

For the ripe, . . .	62 shillings.
" " raw, ...	64 "
" " green, . .	61 "

The produce in money per acre, including straw, stood as follows:

	£.	s.	d.
Ripe, ...	12	17	3½
Raw, ...	13	7	3½
Green, ...	11	11	10¼

The same gentleman made further experiments in 1842, and derived the following results: He cut grain fully ripe, two days before ripe, two weeks, three weeks, four weeks, which specimens are numbered 1, 2, 3, 4, and 5, beginning with the greenest:

100 lbs.	No.	1,	gave	flour	75 lbs.,	shorts	7,	bran	17.
"	"	"	2,	"	"	76 "	"	7,	"	16.
"	"	"	3,	"	"	80 "	"	5,	"	13.
"	"	"	4,	"	"	77 "	"	7,	"	14.
"	"	"	5,	"	"	72 "	"	11,	"	15.

Here No. 3, which is called raw, and was cut two weeks before ripeness, gives 8 per cent. more flour than No. 5, cut ripe. The ripe gave least of all; and No. 4, cut two days before ripeness, the next largest quantity.

Of course the bran and shorts of No. 3 were least, and the bran was found to be thin and soft, while No. 5 was coarse and harsh. The average yield per acre of these experiments was 28 bushels; and the weight of flour in equal measures of wheat was 15 per cent. in favor of the raw over the ripe; the gain in straw, 14 per cent. There was a gain in value of 163 pounds of wheat per quarter, and in the acre yielding 28 bushels a gain of 583 pounds.

INDIAN HISTORY:

Their Relations to us at the time of the American Revolution.

[Western Literary Journal and Review, January 1845.]

THE British authorities in Canada had, during the progress of the American Revolution, solemnly granted the western domain to the Indians residing upon it.

In July, 1744, the Six Nations made a deed of all the lands that were then, or should afterwards be, within the chartered limits of Virginia, to the king of England.*

Notwithstanding the numerous charters which the crown of England had granted to her citizens, companies, and colonies, overlapping each other, and in some instances covering the soil with many thicknesses of paper title, the ministry, at the peace of 1763, undertook to confine all previous grants to the waters of the Atlantic, or to the Alleghany ridge.

There was much apparent force in this position. Neither the king of England, nor the ministers or the people, regarded the vast regions behind the mountains as of much consequence to the nation or her colonies. The home government did not anticipate what Gest and Washington plainly foresaw, that such a soil would be occupied, unless prevented by force. England was ready to give it in exchange for other lands on the north; but the French, confident of success in the field, refused to accept the mountain

* Bancroft's Ab. Vol. II, p. 305.

boundary, which was to branch off in Pennsylvania, pass through French Creek, and include the eastern shore of Lake Erie.* The ruling authority of Great Britain, in the early part of the old French war, appears to have viewed the indefinite regions of the Mississippi as destined to remain a savage wild to the end of time. It was therefore of little consequence, whether it was held by the French, Spaniards, or any other nation, provided such nation should be at peace with England. The French still held Lake Champlain and its vicinage. The English regarded the possession of its shores, and the southern border of Lake Ontario, as a matter of importance; with the eastern slope of the Alleghenies, and the plain between them and the sea, and with the Atlantic waters of New England and Nova Scotia, and the regions of Northern New York and Vermont, they would have been content.

They had not, like the French, seen the latent resources of what is now styled the West. To them, its value consisted merely in the number of skins it would furnish her traders. The thought of introducing civilization into the dark recesses of the Wabash and the Illinois had not possessed an English brain; or if it had, no such project was entertained at the English Court. France had a religious order, whose zeal spread throughout her government. England had her schemes of discovery and conquest; but they were based, not upon an extension of religion or intelligence, but the advantages of commerce. Her citizens, imbibing the same sentiments as her statesmen promulgated, found no sufficient inducements to occupy, or even to explore, the western forests. They were consequently ignorant of the face of the country,

* American Annual Register, 1825-6, and Laws of England, Vol. VI, page 394; Douglass' Summary, Appendix, page 42.

the number and character of the aborigines, the broad and wonderful rivers which flowed past their wigwams, or the soil, unsurpassed in fertility, which exceeded in quantity all that portion of the continent, occupied by the English emigrants. But with Frenchmen it was different; and in 1750, every important stream, whose sources lay in the nooks of the Alleghenies, or whose waters mingled in the northern lakes, had floated the trading barque, cross, and colors of France. She was therefore well acquainted with its unspeakable resources, and could not think of abandoning it without a struggle. The two nations continued to fight, and fortune decided that France should yield to her rival; the lakes and the left bank of the Mississippi.

Still the ministers of Great Britain do not appear to have given credit to the descriptions of Joutel, La Salle, Hennepin, and Charlevoix, respecting the amazing fertility of Canada and Louisiana. They very logically drew the conclusion, that it of right belonged to the crown in some sense, separate from the chartered colonies. Their course of reasoning was thus—when our former kings professed to give away the lands between the Atlantic and Pacific Oceans, the western boundary was remote, unknown, and indefinite: and for this reason alone, an alteration might be permitted, upon later information, changing the nature of the grant.

Although we supposed the British dominions extending across the continent in 1609, and have always claimed that such is the case, yet, long before we acquired actual possession beyond the mountains, our neighbors and competitors had stamped the name of Louis XIV. on the trees of a thousand rivers. Had you, citizens of the new world, been as zealous christians, or as keen discoverers, the banner of St. George would have stood in the place of the Lily and the Cross. We must admit, therefore, that although we possessed the waters of the Hudson, Delaware, and Potomac, we did not know whether the South Sea washed the western base of the mountains from which they sprang, or of other and more distant ranges, until those Catholic missionaries told us it did not. If the occupation of the Connecticut and Susquehanna at the mouth gave us the territory over which they spread their hundred branches, even to the rising springs which the earth spouts forth, then Champlain, Marquette, and La Salle, by planting posts and settlements on the St. Lawrence, and the great river of the West, invested the French crown with an equally good title to the regions drained through their channels.

That being acknowledged, your grants were, in part, the property of another nation at the time, and not in our power to yield. And more, the company whose claim is clearest, embracing the plantations of Virginia, has been deprived of her charter, and the unappropriated lands reverted, in 1624, to the crown. Our own title and that of the French are now united, and the country is at the disposal of King George III.

Such were the premises upon which England assumed the disposal of the western lands, after the treaty of 1763. This was made manifest by a formal proclamation soon after.* In this document, the British government declares its intention to reserve the whole West, for the use and permanent occupation of the aborigines then in possession. All other persons are forbidden to remain or settle within this region; and thus the most civilized nation of the earth decreed the continuance of barbarism over the best portion of North America.

* American Register, 1825–6; Appendix, page 44.

This determination could not have arisen from a sense of obligation to the Indians; for the Hurons, Miamis, and Shawnees, then occupying the Ohio and Lake Erie, had been leagued with the French against them. It was possibly a measure of expediency and humanity, adopted for two reasons—the establishment of peace, and the removal of the Six Nations from New York. Secure in the interposition of a range of mountains, they doubtless considered the abandonment of the region beyond, an easy purchase of the future tranquillity of the Atlantic settlements. Here the aborigines might roam in quiet possession of their hunting-grounds, in a wild, where no white man would desire to enter and abide.

The desires of the home government were for a time gratified by the colonists. But soon after, Daniel Boone, John Finley, John Stewart, Joseph Holden, James Mooney, and William Cool, passed the Cumberland Gap, and reached the Kentucky River. Thus, in the summer of 1769, these roving woodsmen found the country east of the mountains too straightened for them, and came to occupy that far-off wilderness which, six years before, the British statesmen imagined would never be disturbed by the sound of the axe. The assignees of William Penn had also settled themselves around Fort Pitt, with a Gothic determination to remain. Not only the Pennsylvanians, but the Marylanders and Virginians, who accompanied Braddock and Forbes in the expeditions against Fort Du Quesne, had seen the flow of the rivers and the luxuriance of the lowlands beyond the mountains. Those of them who survived to meet their neighbors in the East, did not fail to dwell with eloquence upon the richness of that distant wild, when gathered about the firesides of their homes.

The Virginians began to creep cautiously over the

Blue Ridge, carrying the surveyor's chain and the rifle. They found on the Kanawha and the Ohio some rich bottoms, the counterpart of the James River tobacco fields. The Indians regarded this as an invasion. The men bore the deadliest weapons in their hands; and more, they busied themselves in marking the trees with their axes, and measuring over the country with a mysterious instrument. This looked like war and possession. The western Indians, if they knew of the deed of the Iroquois in 1744, knew also that they did not own the land they pretended to convey. No white sachem had obtained permission to send his young men into their hunting-grounds, to lay them out in parcels, drive stakes, write down memoranda on paper, and to erect a stockade or place of defence.

His untutored mind, by degrees, took in the plan of the Virginians, but his tribes were scattered all along the Ohio, the Scioto, Wabash, and Miami, and could not at once make an official remonstrance to their father, the king of England. His dark eye kindled with anger, as he watched the advance of the white man. The frontier man, in his hunting-shirt and moccasins, had planted himself there, with a determination to make the occupation a matter of life and death. He neither feared or regarded the blanketed vagabond, whom he encountered from time to time. They were bold men who were ready to assert possession by mortal combat. Quarrels and murders were inevitable. The red man knew white men only as one band, conspired against his country. When Michael Cresap at Redstone, and Daniel Greathouse at Wheeling, had secured themselves in Forts, he would pass slyly by those advanced stations, and patiently climb the peaks at the heads of the Kanawha, to strike some unsuspecting settler in the western country. If a horse was seen grazing in the field, he was pressed into the service, and being loaded with plunder, was led back to their towns across the Ohio.

In the guilt or innocence of the border difficulties, the colonial government involved itself as well as its citizens. The House of Burgesses, which sprang into existence under the London Company, had gradually acquired the substantial control of the province. When the French war commenced, in 1754, the colonies, Virginia in particular, entered into the designs of the crown, and freely raised troops and money for the expeditions in the West.

In the poverty of that colony, it became necessary to offer lands in payment of Treasury Warrants, and for military bounties and services. No restrictions were contained in those grants, and the holders were therefore at liberty to spread them upon the soil of any western water within the boundaries claimed by Virginia. These were, as we shall remember, almost without limit on the west and north-west. The crown could not consistently resist these entries, even if they were desirous to protect the Indian in his wild home. The forces, raised by the Virginians, had fought at her (England's) request against her ancient enemy, and conquered for her benefit. They were consequently suffered to enjoy the fruits of the toil, exposure, and danger, that had surrounded them in these campaigns. Such were the circumstances which predisposed the two races, occupying opposite bases of the Laurel Hill, to mutual robbery, retribution, treachery, torture, and death.

Another great event was already (1770) in visible approach. The colonies had been animated by a bold spirit of independence, during an entire century. They had by means of a representative assembly, by continual, but almost imperceptible acquisition, assumed such a control of their political affairs, that the spirit of republican freedom may be said to have been firmly established in America, at the close of the French war.

Just as the whites were attempting, in defiance of the Proclamation of 1763, to plant themselves on the waters of the Ohio, the spirit of liberty, not satisfied with the substance, determined to achieve the name and insignia of independence. Occasions were not wanting to give an air of justice to this determination: indeed, all people are empowered at all times to assert their freedom, and, either with or without cause, to change their form of government.

The agitations of the struggle for an avowed national independence, not only deprived the British of the power to enforce their projects in regard to the Indians, but led the ministers and generals of England to encourage the ferocity of the red man against his white brother. What more was necessary to stimulate both parties to deeds of cruelty, injustice, and revenge? Can the imagination invent a case, where the untamed passions of men might be more inflamed?

This limited review of Indian affairs is given, in order to elucidate the moral relations of that people to the whites, at the time of the Revolution.

INDIAN TITLES IN OHIO.

[Democratic Monthly Magazine, May 1844.]

THE English acquired their first real interest in western soil, as against civilized nations, by the treaty of 1763, which was based upon conquest. In September 1726, the Iroquois ceded to the English crown all their lands west of Lake Erie; but to these territories their title was only nominal, and of little or no value. At the same time and place, Albany, in New York, they granted a tract, sixty miles in width, along the south shore of Lake Ontario and Lake Erie, from Oswego to the Cuyahoga River.*

Afterwards, at Lancaster, in Pennsylvania, in July 1744, for the consideration of four hundred pounds sterling, the Six Nations conveyed "all the lands that are, or may be, hereafter, within the Colony of Virginia."

It was forty years afterwards,† when the independence of the United States had been acknowledged, that the American Commissioners, Oliver Wolcott, Richard Butler, and Arthur Lee, held a council with all the tribes of the Six Nations, at Fort Stanwix.

These nations relinquished all claim to the country west of a line beginning at Johnson's landing, four miles east of Niagara, on Lake Ontario (by them called Lake Oswego), thence southerly, and always

* 2d Bancroft, 236. † October 22, 1784.

four miles distant from the carrying-path between the lakes, to the mouth of Buffalo Creek; thence south to the Pennsylvania line; with it westward to her west line, and south to the Ohio River.* The other tribes do not appear to have contested the ownership of the Six Nations to the eastern end of Lake Erie, as far west as the Cuyahoga River. By the peace of 1783, England assigned all her rights to the United Colonies; whether derived from the French, or the Indians; whether acquired by treaty, or by conquest.

By the treaty of Fort Stanwix, the United States, as a new political body, united in themselves the title of all claimants, civilized and savage, to the north-eastern part of Ohio. It is apparent, that the conveyances, known by the name of treaties, between the French and English, the English and Americans, and also of the Americans and Indians, are based upon conquest. The possession of the country, acquired by war and force, against the will and resistance of the occupant, is the substantial title. The treaty is the evidence of its extent; procured by the successful from the defeated party, upon such conditions as the victorious nation deemed it necessary to impose, or politic to accept.

This is particularly the case with Indian treaties. From the peace of Paris to the American Revolution, the English crown exacted nothing from the north-western Indians but peace, and a restoration of prisoners.

At the close of the Revolution, these tribes, having entered into a war-alliance with our enemies, naturally suffered with them the effects of lawful conquest. Nothing else would have extorted the treaty of Fort Stanwix from the Iroquois, to whom a lasting peace was thus secured.

* American State Papers, Indian Affairs, Vol. I, Page 10

The United States, as we shall soon discover, continued its negotiations with all the neighboring tribes, on the west and north. In these conferences, it is true, matters bore the appearance of bargain and sale; a certain amount of goods and silver for a given tract of land; but without the previous conquest, such councils would not have been held, or such propositions received.

In their *talks*, the Government Commissioners took the position of conquerors treating with an antagonist, whom fortune had abandoned to defeat. The Indians so regarded it, and viewing these treaties as made under durance or over-ruling fear, they assented to them in order to obtain a temporary relief, from policy and necessity.

They never regarded them as contracts between parties free and equal, and, therefore, seldom entered into them as measures to be kept in good faith. From 1784 to 1795, they appear to have made the treaty system a part of the strategy of war.

By the treaty of Fort Stanwix, the Oneidas and Tuscaroras were secured in the possession of their lands, as then occupied.

Having concluded this negotiation, Mr. Butler and Mr. Lee proceeded to Fort McIntosh, on the Ohio, at the mouth of Big Beaver, and were joined by Mr. George Clark, who acted with them in the place of Mr. Woolcott.

There were present delegates and warriors from the Chippewa, Ottowa, Delaware, and Wyandot nations.

The Delawares and Wyandots agreed to confine themselves within the following territory: Beginning at the mouth of the Cuyahoga, where the Six Nations ended; thence up that stream to the "Portage path," and with it to the Tuscarawas branch of the Muskingum; down the same to the forks at the crossing place above Fort Laurens; thence to the mouth of the creek, where the English fort was situated on the Big Miami, and which was taken by the French, in 1752; and thence, with the Portage path, to the Maumee or Omie (St. Mary's River), and with it to the lake, and down along the shore to the Cuyahoga. The lands east, south, and west of these boundaries, they admit to be the property of the United States.

From the Chippewa and Ottawa country, a reservation of six miles square was made at the mouth of the Maumee; a strip six miles on the west side of the Detroit River, as far north as Lake St. Clair, and twelve miles square at Mackinaw.

A mutual peace was concluded with them all; the Wyandots delivering three hostages, and the Delawares two, as a guaranty for their engagement to return all prisoners, white or black *—concluded January 21st, 1785.

The Shawnees were not regarded by the other tribes as owning any territory in the north-west. They were, however, so willing and efficient in waging war upon the whites, that a spacious and rich country was allotted them upon the Scioto.

A council was held with them at the mouth of the Great Miami, in January, 1786, by Messrs. Butler and Clark, assisted by S. H. Parsons, Esq. Here they acknowledged the United States to be sovereign within the territory relinquished by Great Britain, and promised to deliver up all prisoners. By consent of the other tribes, they were allotted as a hunting-ground a tract on the head-waters of the two Miamis, about forty miles broad, running westward from the Wyandot and Delaware boundary to the Wabash River—concluded January 31st.

* State Papers, Vol. I, page 11.

Three years after this transaction, General Arthur St. Clair convened the Six Nations at Fort Harmar, at the mouth of the Muskingum.

The United States were making such rapid acquisitions of the Indian territory, that the Mohawk chief Thay-an-da-negea had, in the meantime, like Pontiac, concocted another confederacy, which will receive a separate consideration.

The five nations, the Chippewas and Wyandots, the Ottowas, Delawares, Shawnees, Twichteewees or Miamis, Potawotomies and Cherokees, together with the confederation of the Wabash tribes, assembled in convention at the Huron village, near the mouth of Detroit River. This imposing council joined in an eloquent address to the Congress of the United States, declaring all treaties void that were not sanctioned by all the tribes. This production is dated December 18, 1786.

But this did not prevent the Six Nations, with the exception of the Mohawks, the tribe of which Brant was chief, from treating with General St. Clair at Fort Hamar.

In consideration of $3,000 and certain presents, they confirmed the treaty of Fort Stanwix, and agreed to deliver up horse-thieves, and murderers, to be punished by our laws. The Mohawks had the privilege of assenting to the treaty, in case they should desire to become parties.

At the same council-ground appeared the Wyandots, Delawares, Chippewas, and Ottowas, who solemnly confirmed the treaty of Fort McIntosh. Also, the Potawotomies and Sauks, who relinquished their rights to the country east, south, and west, of the old Delaware and Wyandot boundary, and a reserve of six miles square at Fort Sandusky. An universal agreement was made to sell to no power, or person.

but the United States; and to deliver up all murderers and robbers. They were permitted to inflict such punishment upon intruders on their lands as the several tribes should see fit. The amount here paid the western Indians was $6,000 in money, together with liberal provisions and certain presents. A complaint was entered by the Wyandots against the Shawnees, whom they said were troublesome, not only to the United States, but to the other tribes, and were merely living by sufferance among them.

They threatened that, unless the Shawnees would be at peace, they should be dispossessed of the hunting-ground described at the mouth of the Miami, in 1786, and affirmed that this region, of right, belonged to the Wyandots. This treaty was finally assented to on the 9th of January, 1789; and the government regarded the title to the State of Ohio, south of the Fort Loramie's and Fort Laurens line, and east of the Cuyahoga, as settled beyond question.

During the previous summer and fall, white settlements had been formed at the Muskingum and the little Miami, and also at Cincinnati.

Public surveys had been commenced north-west of the Ohio, soon after the treaty at Stanwix; and had now extended over large tracts of country.

The mass of the Indians were still fixed in the determination that the Americans should remain south of the Ohio. The surveys, and finally the permanent settlements, so rapidly forming on its northern bank, exasperated them to madness.

Treaties, signed by chiefs and head-men, had little influence in restraining young men and warriors, who thirsted for the blood of the whites, and murders were continually perpetrated.

The new government of the United States, under the Constitution, soon determined to make serious war upon them; but its resources were limited, and its generals unsuccessful. It was not until after the terrible infliction, at the battle of the Rapids, in 1794, that the Indian concluded to yield to his destiny.

The consequence of this action was the ratification of a new treaty, at Greenville, on the 3d of August, 1795. It was signed by the Wyandots, Delawares, Shawnees, Ottowas, Chippewas, Pottawotomies, Miamis, Eel-River-Indians, Weas, Kickapoos, Piankeshaws, and Kaskaskies; twelve tribes, who had suffered in the engagement the year before.

In 1793, when Colonel Hardin and Major Trueman were sent into the Wabash country, with offers of peace, they were taken as prisoners and murdered. Other offers were made, by General Wayne, in 1794, before the battle, and had been refused. They were repeated after the victory, but the chiefs and warriors were slow to assent to the formation of a new treaty, which, they were well aware, must be severe upon them. They returned to their villages to consult their tribes, and to confer with the British, who still held possession of the northern posts, and prompted them to the war.*

But at length they were assembled before General Wayne and his troops, at Fort Greenville, and agreed upon the following terms of peace, and boundaries. The old Wyandot and Delaware boundary was re-affirmed, from the mouth of the Cuyahoga to Fort Loramie; from thence it was continued to Fort Recovery, situated on the ground of St. Clair's defeat, in 1792; and thence to the Ohio River, opposite the mouth of the Kentucky River. With the exception of some reservations for forts, in the north-west, the United States relinquished to the Indians all claim to the territory north and west of this line.

* Major Hamtramck's Letter Book, 1795, Forts Wayne and Defia.

The Indians contracted not to sell to any power but the United States, and received $20,000 in goods. An annuity of $9,500 in goods was promised them, which the government regarded as more beneficial to them than money. The reservations in Ohio were, a tract, six miles square, at Fort Sandusky, on the bay of Sandusky; six miles square at Fort Defiance; six at Girty's town, at the north end of the Miami and St. Mary's Portage; and six at the mouth of Loramie's Creek on the Miami. If anything should be found in treaties, made since 1783, conflicting with the terms of the present treaty, such parts were to be considered void.* Ten hostages were taken, to secure the restoration of all prisoners, and the nations remained at peace until Tecumseh arose among them, and reorganized the confederacy, in 1811.

We come next, in the order of time, to the treaty of purchase, made at Fort Industry on the Miami of the Lakes, July 4, 1805, by Charles Jouett, on the part of the United States, Henry Champion as the representative of the "Connecticut Land Company," and J. Mills for the "Sufferers' land." The Indian tribes represented, were the Wyandots and Delawares, the principal owners; the Chippewas and Ottowas, Shawnees, Munsees, and Pottawotomies.

For the consideration of an annuity of $1,000, and the payment in hand of $18,916, they ceded to the United States a tract lying south of the Reserve and the Sufferers' land, and west of the Tuscarawas, to the Greenville treaty line, bounded on the west by the meridian of the west line of the "Fire" or "Sufferers' lands." They relinquished to the Connecticut Land Company all the land between the Cuyahoga River and the Fire lands, or "Sufferers," being the remainder of the Reserve.

* State Papers, Vol. I, p. 595.

There remained, therefore, only that portion north of the Greenville line, and west of the meridian of the Reserve, to which the Indian title attached.

A portion of this was purchased by General William Hull, at Detroit, on the 7th of November, 1807. At that time, the Ottowas, Chippewas, Wyandots, and Pottawotomies, granted as follows: from the mouth of the Miami of the Lakes, up its channel, to the mouth of the Auglaize; thence north to the latitude of the south end of Lake Huron; thence north-east to the "white rock," on its western shore, and with the territorial boundary to the mouth of the Maumee.

They received in return $10,000, and two "white blacksmiths," to reside among them.*

On the 25th of November, 1808, General Hull perfected another purchase, of the Chippewas, Ottowas, Pottawotomies, Wyandots, and Shawnees, at Brownstown, in Michigan, by which they granted a strip two miles in width, for the purpose of constructing a road, from the rapids of the Miami to the west line of the Reserve; also, one hundred and twenty feet in width, for a road from Lower Sandusky to the Greenville line.†

The old grudges of 1786 were now reviving in the breasts of the northern and western tribes, on account of the rapid increase and acquisitions of the whites. The British, in anticipation of a war with the United States, made use of this hostile inclination against us, and the war of 1812 ensued.

The main body of the Wyandots, Delawares, Senecas, and Shawnees, remained faithful to the United States. They had engaged, during the war, against the Miamis of the Wabash, on their own responsibility. This warfare was adjusted by Generals Harrison and

* State Papers, p. 747. † State Papers, p. 759.

Cass, at Greenville, July 22, 1814, and these tribes agreed to a mutual peace.*

Soon after the peace with Great Britain, other treaties of purchase were made. At one, dated "Spring Wells," near Detroit, September 18, 1815, the Chippewas, Ottowas, Pottawotimies, and those bands of the Wyandots, Delawares, Senecas, Shawnees, and Miamis, who had been with the British during the war, confirmed and renewed the treaty of Greenville, made by General Wayne, in 1795.

The Miamis had repented before the close of the war, and had signed a treaty at Greenville, in 1814. The chiefs and warriors of the other tribes present, who had taken up arms against us, were pardoned, and their possessions restored. This negotiation was conducted, on the part of the United States, by Generals Harrison and McArthur, and by John Graham, Esq.

The tribes were again assembled by Generals Mc Arthur and Cass, at the Maumee Rapids, on the 29th of September, 1817.

The Wyandots, computed at one thousand persons, ceded all their lands west of the meridian of the Reserve, north of the Greenville line, to Loramie's; east of the Portage path and the St. Mary's River to Fort Wayne; thence down the north bank of the Maumee to the meridian of the treaty of Detroit; thence down the middle of the Maumee to the mouth of the Auglaize, and so with the channel, as described in the treaty of Detroit, in 1807, to the lake, and thence to the meridian of the Reserve. They received in return an annuity of $4,000 in specie, and certain reservations.

The Pottawotimies, Ottawas, and Chippewas, relinquished a tract on the north of the Maumee, be-

* State Papers, p. 826.

tween the meridian of the west line of Ohio and the meridian of the treaty of Detroit, measuring forty-five miles on the latter meridian northward.

The Pottawotimies were granted an annuity of $1,300 for fifteen years; the Ottowas of $1,000 for the same length of time; and the Chippewas in like manner, $1,000.

The former annuities to the Wyandots, Shawnees, and Delawares, in goods, were changed into specie, and those due the Ottowas and Chippewas, by the treaty of Greenville, were likewise reduced to cash payments.

The Delawares were to receive $500, in 1818; the Senecas an annuity of $500, and the Shawnees of $2,000; principally in consequence of their adherence to the government during the war. The friendly tribes were paid for losses the sum of $14,478.

At the same time, the American Commissioners reconveyed to the Wyandots twelve miles square, at Upper Sandusky, having Fort Ferree in the centre, and also thirty thousand acres on the Sandusky River.

To the Shawnees, ten miles square, with the centre at Wapahkonnetta, and twenty-five square miles on Hog Creek.

To the Shawnees and Senecas, forty-eight square miles near Loramie's.

To the Ottowas, five miles square, where the trace or trail crosses Blanchard's Fork of the Auglaize.

Also, various small grants to meritorious persons and chiefs, as per schedule.

The Delawares received nine miles square on the Sandusky, including Captain Pipe's village; and the Ottowas thirty-four square miles, including McCarty's village, on the south side of the Maumee.*

* State Papers.

The sum total of these grants is three millions, five hundred and eight thousand, seven hundred and sixty acres; and of the Indian Reserves, one hundred and eighty-five thousand, seven hundred and eighty acres; making three millions, six hundred and ninety-four thousand, five hundred and forty acres in all. The share conveyed by the Wyandots was three millions, two hundred and thirty-one thousand, five hundred and sixty acres, and the amount paid to all the tribes, reduced to present cash payments, $140,893, or three cents and eight mills per acre.*

The remaining corner of Indian territory, between the St. Mary's, the Portage path, and the Greenville line, was secured to the United States by cession, from the Miamis at the treaty of St. Mary's, October 6th, 1818. The Commissioners, on the part of the United States, were General Lewis Cass, Jonathan Jennings, and Benjamin Parkes.†

The Reserves allotted to various tribes in Ohio have been sold by them from time to time, closing with the twelve miles square reservation at Upper Sandusky, which was ceded by the Wyandots in 1842.

There were treaties of amity and peace, in addition to those here enumerated, in which nothing is said concerning the acquisition of territory. At the reduction of Fort Du Quesne, in 1758, by General Forbes, the Indians, engaged with the French against the English, concluded a treaty of peace. The northwestern Indians did the same at German Flats, in December 1764, the result of Bradstreet's expedition against them. In May of the following year, the Ohio Indians, who had met Colonel Boquet on the Muskingum, made a treaty with Sir William Johnson. Lord Dunmore, in November 1774, also concluded an amicable arrangement with them at Camp Charlotte,

* State Papers, Vol. II, p. 131. † Laws of 1819.

in the county of Pickaway. The authorities of Pennsylvania concluded a compact of amity with the adjacent tribes, in 1765.*

It was thus only after the execution of fifteen treaties of peace and purchase, during the period of sixty years, by means of eleven principal military expeditions into their country, not including those of the war of 1812, and after fighting seven important engagements within our own State, at a loss of about twelve hundred men, that our soil was finally relieved from the presence, and our citizens from the dread, of the Indian race.

From the outbreak of the French war, in 1754, to the battle of the Rapids, in 1794, forty years, it is estimated that five thousand perons, men, women, and children, suffered captivity or death, on the western frontiers.

The number of whites slain in battle exceeded the number of warriors destroyed by private and public efforts.

Yet, in 1811, the five tribes remaining in Ohio, could muster only about two thousand warriors, or eight thousand souls in all. In 1764, they were estimated at three thousand warriors, and fifteen thousand in the aggregate, the proportion of women and children being greater during peace, and before their difficulties with the whites.

* Craig's Discourse on the Boundary of Pennsylvania.

www.ingramcontent.com/pod-product-compliance
Lightning Source LLC
LaVergne TN
LVHW020112110826
845151LV00001B/136